TRANSPORTING NATURAL GAS FROM THE ARCTIC

The Alternative Systems

TRANSPORTING NATURAL GAS FROM THE ARCTIC
The Alternative Systems

Walter J. Mead
with George W. Rogers
and Rufus Z. Smith

American Enterprise Institute for Public Policy Research
Washington, D.C.

Walter J. Mead is professor of economics at the University of California, Santa Barbara.

George W. Rogers is professor of economics, Institute of Social, Economic and Governmental Research at the University of Alaska at Juneau.

Rufus Z. Smith was deputy assistant secretary for Canadian affairs at the U.S. Department of State.

ISBN 0-8447-3270-2

Library of Congress Catalog Card No. 77-85373

AEI Studies 171

Printed in the United States of America

CONTENTS

PREFACE

1 **U.S. NATURAL GAS RESOURCES AND ALTERNATIVES
FOR THE FUTURE**   **1**
Alternatives for the Future   2
Conservation   4

2 **ALTERNATIVE TRANSPORTATION SYSTEMS FOR
DELIVERY OF PRUDHOE BAY GAS**   **7**
The Arctic Gas Project   7
The El Paso Project   9
The Alcan Projects   10
Estimates of Gas Shrinkage   17

3 **THE ENVIRONMENTAL IMPACTS OF THE
ALTERNATIVE ROUTES** *George W. Rogers*   **21**
Principal Assessments   22
Scope of the Present Assessment   27
Physical Features of the Alternative Systems   30
The Living Environment: Wilderness and
Wildlife   40
The Human Environment: Socioeconomic
Impacts   44
General Observations of the Process of Environmental
Assessment   47
The Final Environmental Assessment   51

4 THE CANADIAN SOVEREIGNTY ISSUE *Rufus Z. Smith*   **53**
   Physical Security   53
   Political Security   55

5 ECONOMIC ANALYSIS OF ALTERNATIVE ROUTES   **61**
   Procedures Followed to Develop Cost and
      Benefit Data   61
   Capital Cost Estimates   62
   Cost-of-Service Estimates   66
   Net National Economic Benefits
      Estimates   75

6 THE COST OVERRUN PROBLEM   **83**
   Maximum Acceptable Cost Overrun   83
   Probable Construction-Cost Overruns   84
   Project-Specific Estimates of Cost Overruns   90
   Summary of Cost-Overrun Analysis   93

7 RATE REGULATION AND FINANCING   **95**
   The Unique Problems of Rate Regulation
      and Financing   95
   Positions of the Parties   96
   The Dilemma of Regulation and
      Finance   98
   The Missing Link   100

8 CONCLUSIONS AND RECOMMENDATIONS   **107**

PREFACE

This book is an analysis of the alternative systems proposed to the Federal Power Commission for transporting natural gas from the Arctic to the lower United States. Because events that will affect the final decision have occurred during the period of writing, a note on the current status of the various proposals seems in order.

Throughout the hearings held by the Federal Power Commission in 1976, the system proposed by Arctic Gas seemed to be the front-runner. Its advantages included delivery of both Prudhoe Bay and Mackenzie Delta gas at a lower cost of service and higher net national economic benefits than the alternative routes. The FPC staff and.the FPC administrative law judge both recommended approval of Arctic Gas over the El Paso and Alcan routes. Then Alcan submitted a modified route designed to meet most of the FPC's objections.

The situation changed dramatically when the Canadian Berger Commission and the Canadian National Energy Board ruled against the Arctic Gas proposal. Since then, the Alcan #2 route via Dawson has become the favored route, with El Paso a close second. The prime minister of Canada has communicated by telephone to President Carter his support for the Alcan route. The President will make his recommendations known within ninety days following September 1, 1977. The final judgment will be made by the Congress of the United States and the Canadian Parliament.

Regardless which route is ultimately approved, important questions of regulation and rate-making must be resolved. These questions are also addressed in this study.

I am indebted to several people for their assistance. Mr. Kemm Farney conducted most of the research evaluating the cost overrun record for twelve major construction projects. I am indebted to Russell Jones and Scott Beam for research and editorial assistance. All are graduate students at the University of California at Santa Barbara. My colleague Professor Phillip E. Sorenson read both drafts of the manuscript and offered valuable advice. Also, conversations with George Rogers and Rufus Smith were helpful at many points. Others too numerous to mention provided data and comments. Responsibility for this report rests entirely with the principal author.

WALTER J. MEAD
August 1977

1
U.S. NATURAL GAS RESOURCES AND ALTERNATIVES FOR THE FUTURE

Natural gas is a nonrenewable resource. Therefore, it should surprise no one to learn that past growth rates in proved gas reserves and annual gas production cannot possibly be sustained. From the end of World War II, natural gas production in the United States increased at a compound annual rate of 5.8 percent reaching a peak in 1973. As new frontiers were explored, proved reserves increased at a compound annual rate of 3.2 percent, ending in 1967.

The Prudhoe Bay oil and gas field was discovered in 1968. The largest known reservoir of either oil or gas in North America, this field is estimated to contain 9.6 billion barrels of crude oil and 26 trillion cubic feet of natural gas. It covers an area of about 200 square miles on the North Slope of Alaska, bordering on the Beaufort Sea. The addition of Prudhoe Bay increased U.S. oil reserves in 1970 by 32 percent and natural gas reserves by 9.5 percent.

Energy specialists have known for several years that conventional supplies of natural gas available to the U.S. market could not continuously be expanded to meet past growth rates but would inevitably decline. Reserves of natural gas in the United States reached a peak in 1967 of 293 trillion cubic feet (Tcf). Even with the addition of Prudhoe Bay, reserves declined to 220 Tcf by the end of 1976 (see Table 1). With this decline in available reserves, it was inevitable that the production of gas would also peak out and then decline. Production reached a peak of 21.7 Tcf in 1973 and declined 12 percent to 19.0 Tcf in 1976.

Prudhoe Bay reserves of natural gas are expected to support production at about 2.4 billion cubic feet per day (Bcf/d) or 876 billion cubic feet per year (Bcf/yr). Thus even with the addition of Prudhoe Bay, annual U.S. natural gas production will be far below the 1973 peak.

At the same time that U.S. domestic production of natural gas was declining, imports from our traditional foreign supplier, Canada, also

1

Table 1

NATURAL GAS RESERVES AND PRODUCTION,
UNITED STATES AND CANADIAN MACKENZIE DELTA,
1967-76

Year	U.S. Excluding Prudhoe Bay		Prudhoe Bay		Canadian Mackenzie Delta	
	Reserves (in Tcf)	Production (in Tcf/yr)	Reserves (in Tcf)	Production (in Bcf/d)	Reserves (in Tcf)	Production (in Bcf/d)
1967	293[a]	17.4				
1973	250A[b]	21.7[a]				
1976	220A[b]	19.0	26[c]	2-2.5[c]	4.3-5.1[c]	1.0[c]

[a] Peak figure.
[b] Includes Prudhoe Bay.
[c] Estimated.

Source: American Gas Association, *Reserves of Crude Oil, Natural Gas Liquids and Natural Gas in the United States and Canada as of December 31, 1975*, p. 120; Bureau of Mines, *Minerals and Materials*, June 1977, p. 13; American Gas Association, Press Release, April 7, 1977; United States Department of the Interior, *Alaskan Natural Gas Transportation Systems*, December 1975, p. 4; Federal Power Commission, *Recommendation to the President, Alaska Natural Gas Transportation Systems*, May 1, 1977, p. III-51.

declined. Imports peaked out at 2.83 Bcf/d in 1973 and declined to 2.65 Bcf/d in 1976.

In view of these deteriorating reserve and supply conditions following a history of expansion, new gas supplies, either as an alternative to Prudhoe Bay gas or as a supplement to any future Prudhoe Bay production, must be considered.

Alternatives for the Future

Increased Supplies from Canada. It is extremely unlikely that U.S. imports of natural gas from Canada will increase. Domestic demand for natural gas in Canada has been growing faster than Canadian production, and domestic markets are served first. Exports declined from a peak of 2.83 Bcf/d in 1973 to 2.65 Bcf/d in 1976. A study prepared by the Canadian National Energy Board (NEB) forecast that Canadian domestic demand would equal domestic production (excluding any possible production from the frontier areas—the Mackenzie Delta, the Beaufort Sea, and the Arctic Islands) by 1984, after which a domestic deficit would grow.[1] In the absence of new supplies from frontier

---

[1] National Energy Board, *Canadian Natural Gas Supply and Requirements,* April 1975, pp. 21 and 59.

areas, the exportable surplus must decline to zero by early 1985 or shortly thereafter.

Liquefied Natural Gas Imports. Liquefied natural gas (LNG) may be available in large quantities from Indonesia, Algeria, Nigeria, and Middle Eastern countries. The most recent LNG imports authorized by the Federal Power Commission (FPC)—Truckline LNG imported from Algeria—permit a landed price of $3.37/million Btu. At 1,150 Btu/cubic ft., this translates into $3.88/Mcf. The city gate wholesale price would be about $3.60/million Btu ($4.14/Mcf).[2]

Natural Gas Produced from Devonian Shales in the Appalachian Region. In the eastern part of the United States, large-scale natural gas resources are known to exist in Devonian shales, primarily in the Appalachian plateau. With the price of natural gas kept low by federal controls, very little production has taken place in these fields. Using conservative assumptions, the Office of Technology Assessment has estimated that at wellhead prices for natural gas in the $2.00 to $2.50/ Mcf range, approximately 1 trillion cubic feet of gas could be produced annually over the next fifteen to twenty years from readily recoverable reserves.[3] Larger quantities would be producible at higher prices, of course.

Synthetic Gas from Coal. The huge reserves of coal in the United States offer another alternative supply of gas. Synthetic gas can be made from coal by processes that have been available for several decades. New research in coal-gas technology may also result in cost reducing breakthroughs. Using existing technology, synthetic gas may be produced from coal in the post-1990 years, the cost ranging from a low of $3.50/Mcf to a high of $4.50/Mcf (constant dollars, wholesale).[4]

Synthetic Gas from Oil. Synthetic gas may also be produced from oil. The cost of this source of gas currently is estimated to range from $3.50 to $5.50/million Btu.[5] Apart from its high cost, synthetic gas from oil is not an attractive alternative because it would increase U.S. dependency on imported oil.

Other Sources. Large supplies of gas will be available in the more distant future from resources currently known to exist. These include gas from tight formations, geopressure zones, and hydrates. These

---

[2] Federal Energy Administration, *National Energy Outlook,* February 1976, p. 157.
[3] Office of Technology Assessment, *Enhanced Recovery of Oil and Devonian Gas,* June 1977, part II, p. 5-6.
[4] White House Task Force, *Report of the Working Group on Supply, Demand and Energy Policy Impacts of Alaska Gas,* July 1, 1977, p. 121.
[5] Ibid., p. 109.

sources are discussed in a recent United Nations publication.[6] While potential supplies from these sources are very large, cost data are not available.

This brief survey of alternative natural gas supplies suggests that only imported LNG could supply large quantities of moderately priced natural gas for use in the near future. The alternative large-scale *energy* sources for the remainder of this century consist of enormous U.S. coal reserves, increased oil imports, and nuclear power. According to a recent White House task force report, gas supplies from the sources reviewed will become plentifully available only at city gate prices above $4.00/Mcf.[7]

Conservation

"Conservation" is endorsed by everyone and understood by very few. From an economic standpoint, conservation of resources is attained when their present value is maximized. In other words, the optimum level of resource conservation occurs when the *in situ* value of the resources increases at a compound annual rate equal to the rate of return on other investments of similar risk. Resource conservation decisions should be made in the same manner as other investment decisions. An owner of energy resources seeking to maximize their present value should save them for future generations (leave them in the ground) if he expects their *in situ* value to increase faster than the return on the best alternative use of his scarce capital.

Nothing in the history of federal government energy policy to date inspires confidence that new energy policies will lead to conservation in an economic sense.[8] For most of the last half century, the federal government has promoted a low-price policy for energy and, thereby, high demand levels.[9] The major effect of approximately fifty years of tax policy affecting oil and gas (both the percentage depletion allowance and tax provisions permitting the expensing of intangible drilling costs) has been to increase production and, therefore, lower prices. Moreover, interstate natural gas has been subject to wellhead price controls since 1954 with the effect that prices have been artificially low. In the case of crude oil, price controls were imposed on August 15,

---

[6] United Nations Institute for Training and Research, *The Future Supply of Nature-Made Oil and Gas* (New York: Pergamon Press, 1977), pp. 13-18.

[7] White House Task Force, *Report*, p. 127.

[8] For a brief review of U.S. energy policy, see W. J. Mead, "An Economic Appraisal of President Carter's Energy Program," *Science,* vol. 197 (July 22, 1977), pp. 340-345.

[9] This low-price policy has not been followed consistently. The effect of market demand prorationing and import quotas has been to artificially raise petroleum prices.

1971. The 1975 Energy Policy and Conservation Act rolled back the price of crude oil still further. To the extent that price controls on crude oil have caused product prices to be lower than they would otherwise be, the demand for petroleum products has been stimulated. Domestic supplies have been artificially reduced by that same policy. The growing gap between domestic supply and domestic demand has been filled by sharply rising imports. While conservation is widely applauded, past government policy has not served, and future government policy probably will not serve, the goal of conservation.

2
ALTERNATIVE TRANSPORTATION SYSTEMS FOR DELIVERY OF PRUDHOE BAY GAS

Three different routes have been proposed for transportation systems to deliver gas from the Prudhoe Bay field on the North Slope of Alaska to markets in the lower forty-eight states. Two of these—Arctic Gas and Alcan—are buried all-pipeline routes running through Canada and delivering gas to both the far West and the Midwest. The other proposed route—El Paso—does not cross Canada but requires converting the natural gas into liquefied natural gas and shipping it in cryogenic tankers to a port in southern California.

The Arctic Gas Project

The first project, submitted by a consortium of sixteen American and Canadian companies, primarily gas pipeline companies, proposes a 4,512-mile buried pipeline route (see Figure 1). On March 21, 1974, the Arctic Gas sponsors proposed a forty-eight-inch diameter pipeline running east along the coastal plain of the North Slope for 195 miles to the Canadian Yukon border and passing through Alaska's Arctic National Wildlife Range. From the Alaskan border, the line would run another 177 miles along the Yukon-Beaufort Sea shore to connect with a 19-mile lateral from Canada's Mackenzie Delta gas field. Carrying the commingled Mackenzie Delta and Prudhoe Bay gas, the pipeline would continue up the Mackenzie River Valley for 1,410 miles to Caroline Junction in southern Alberta. At this point, the Mackenzie Delta gas would be diverted to Canadian markets.

Prudhoe Bay gas would flow into the United States through two lines, an eastern and a western leg. The eastern leg would pass through Monchy, Saskatchewan, to serve midwestern markets and would terminate in Dwight, Illinois, near Chicago. The western leg would cross

Figure 1

BURIED PIPELINE ROUTE PROPOSED BY ARCTIC GAS

Source: Federal Power Commission.

into the United States at Kingsgate, Idaho, and would terminate in Antioch, California. Beyond the terminal points, Prudhoe Bay gas would flow through existing pipelines. Markets not entirely supplied by Prudhoe Bay gas would be supplied by gas from fields located mainly in the southern tier of states.

The principal advantages of the Arctic Gas proposal are that it would utilize an all-pipeline system which totally avoids mountainous areas and earthquake zones and that it would tap both Prudhoe Bay and Mackenzie Delta reserves. Its principal disadvantages are that, probably as a precondition to construction, the Canadian government and the Canadian Arctic natives would have to settle their land claims dispute, that it would cross the Arctic National Wildlife Range, and that, for most of the distance from Prudhoe Bay to the Alberta border, it would cross relatively pristine areas. Like the Alcan routes, it has the possible disadvantage of passing through foreign land.

Nahum Litt, the administrative law judge of the Federal Power Commission (FPC), in his decision filed on February 1, 1977, favored the Arctic Gas proposal over two other proposals described below (El Paso and Alcan #1). On May 1, 1977, two FPC commissioners recommended approval of Arctic Gas, while the other two recommended the Alcan #2 proposal.

There was an adverse ruling by Canadian Justice Thomas R. Berger on April 15, 1977, based primarily on Canadian environmental and Arctic native concerns. Then in June 1977, the Canadian National Energy Board ruled that a pipeline along the coast of the northern Yukon was environmentally unacceptable to the board, as was the cross-delta section of that route. These two Canadian decisions, while not final, should be viewed as clear expressions of the Canadian viewpoint. On July 29, 1977, eight American utility company sponsors of the Arctic Gas proposal, reacting to Canadian developments, agreed to shift their support to the Alcan #2 proposal, apparently eliminating Arctic Gas as a viable possibility.

The economic, environmental, and political analysis contained in this report will show data for Arctic Gas, along with El Paso and Alcan, for two reasons: first, for comparison with El Paso and Alcan, and second, because in the world of governmental decision making, dead is not always dead. Circumstances might change in such a way as to revive the Arctic Gas alternative.

The El Paso Project

The second proposal was received from the El Paso-Alaska Company and is designed to avoid any possible problems arising out of passage through Canadian territory (see Figure 2). This proposed route calls for a pipeline forty-two inches in diameter running 809 miles from Prudhoe Bay to Gravina Point on the ice-free coast of southern Alaska (approximately the route of the Alyeska oil pipeline). At Gravina Point, the natural gas would be converted to LNG by lowering its temperature in an LNG plant to minus 259° Farenheit. In the process of liquefaction, the cubic volume of the gas would be drastically reduced, one cubic foot of LNG being the equivalent of 623 cubic feet of natural gas.

The LNG would be shipped from Gravina Point, Alaska, to a California port, possibly at Point Conception, a distance of 1,900 nautical miles (2,200 statute miles). To transport 2.4 billion cubic feet of gas per day, nine cryogenic LNG carriers would be required, each carrying 165,000 cubic meters of LNG. One such tanker would leave Gravina Point, Alaska, every thirty-six hours continuously for approximately twenty years.

In southern California, the LNG would be regasified and transported via pipeline to markets primarily in the western half of the United States. Gas from other fields in the southern tier of states would be redirected, through a process called displacement, to markets farther east.

The principal advantages offered by the El Paso proposal are that it would avoid any possible problems arising out of Canadian national sovereignty, provincial conflict, or native land claims; it would utilize much of the existing Alyeska access and haul-road system; and it would permit delivery of gas to Fairbanks and other areas in eastern Alaska. Its principal disadvantages are that it would involve an expensive and relatively dangerous process of liquefaction, shipment by LNG carriers on the high seas, and regasification. In addition, the gas pipeline would cross an active earthquake zone in the Chugach Mountains of southern Alaska, and the LNG processing facility would be located in an earthquake zone. Finally, it would deliver gas in the southwestern sector of the United States where it is least needed, requiring expensive displacement of existing gas flows.

The Alcan Projects

The Alcan #1 proposal was submitted to the FPC on July 7, 1976, two years after the El Paso and Arctic Gas proposals. Like the Arctic Gas route, the Alcan proposals call for all-pipeline routes running through both Alaska and Canada, delivering gas to both the Midwest and the far West. Both Alcan routes run approximately parallel to the Alaskan

Figure 2

PIPELINE AND SHIPPING ROUTE PROPOSED BY EL PASO-ALASKA

Source: Federal Power Commission.

11

oil pipeline to a point south of Fairbanks and then approximately follow the Alcan Highway into northern British Columbia (see Figure 3).[1] In British Columbia and Alberta, Alcan #1 proposed to utilize two existing pipelines, increasing their capacities through the construction of parallel lines in a process known as looping.

Because of its apparent economic inefficiency and inadequate research support, the Alcan #1 proposal did not fare well before the FPC. The FPC staff's position brief concluded that the Alcan proposal was not economically viable, and the FPC administrative law judge declared that Alcan finished dead last.

There were three primary reasons for this low rating of Alcan #1 by the FPC staff and administrative law judge. First, the proposed forty-two-inch-diameter line from Prudhoe Bay to British Columbia was small and inefficient compared with the forty-eight-inch line proposed by Arctic Gas. Second, while the proposal to utilize two existing pipelines sounded plausible, it required looping which, together with fuel inefficiency, made this phase of the Alcan #1 proposal expensive. Third, because Arctic Gas, with which it was being compared, provided for the transport of both Alaskan and Canadian gas, the Alcan #1 proposal was considered in conjunction with the proposed Maple Leaf pipeline, which would run from the Mackenzie Delta to northern Alberta. Together the Alcan #1 and Maple Leaf lines required high mileage and consequently high costs.

While the economics of Alcan #1 paired with Maple Leaf were unfavorable, the environmental interveners (the Sierra Club, the Wilderness Society, the National Audubon Society, and the Alaska Conservation Society) as well as the FPC environmental staff and the Interior Department staff were favorably disposed to the Alcan route. In view of this fact, the Alcan Pipeline Company on March 8, 1977, proposed Alcan #2. Alcan #2 called for a forty-eight-inch-diameter line with an average daily throughput of 2.4/Bcf.

The two Alcan proposals follow the same route from Prudhoe Bay through Alaska and the Yukon Territory into northern British Columbia. Near Fort Nelson, Alcan #2 abandons both the ideas of following the Alcan Highway and of utilizing existing Canadian pipeline facilities. Instead, it proposes a new forty-eight-inch pipeline across relatively flat country for 400 miles southeast to a point near Caroline Junction, Alberta. At this point Alcan #2 would be divided into two

[1] In three instances the proposed Alcan route runs through mountain passes that differ from those used by Alyeska. Also, an Alcan gas pipeline would have between forty-three and forty-seven crossings of the Alaskan oil pipeline, twenty-six to thirty would be crossovers, while seventeen would be in the buried mode.

Figure 3

PIPELINE ROUTES PROPOSED BY ALCAN

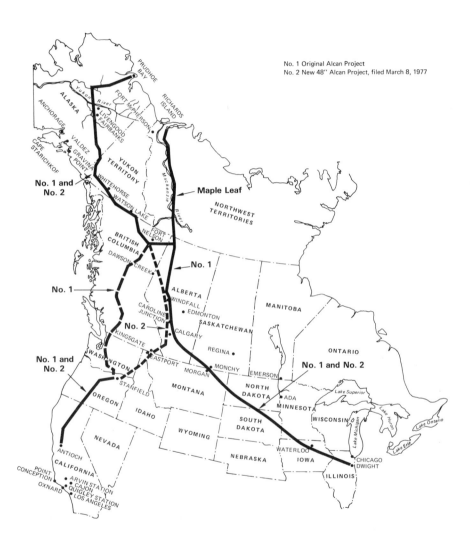

No. 1 Original Alcan Project
No. 2 New 48" Alcan Project, filed March 8, 1977

Source: Federal Power Commission, modified by the author.

13

legs, one forty-two-inch line serving the Midwest and a second thirty-six-inch line serving the far West.

Alcan #2 via Dawson, Yukon, with a Dempster Lateral. The June 1977 report of the Canadian National Energy Board may force a modification of Alcan #2. The NEB report recommends approval of a line running from the Mackenzie Delta to the town of Dawson in the Yukon Territory, via the Dempster Highway. This proposal calls for a thirty-inch, 460-mile pipeline that would deliver Canadian gas primarily to Canadian markets. The NEB report also recommends modifying the Alcan #2 route so that it detours about 100 miles northeast of the Alcan Highway and passes through Dawson (see Figure 4). The modified Alcan #2 route, for which some preliminary cost estimates are available, would leave the Alcan Highway at Tetlin Junction near the Alaska-Yukon border and follow the Taylor Highway into Dawson. From Dawson, the line would follow the Klondike Highway to Whitehorse in the Yukon where it would rejoin the Alcan Highway. This modification would add 120 miles to the Alcan #2 proposal.

The object of the modifications proposed by the NEB would be to permit future delivery of Mackenzie Delta gas to Canadian markets while avoiding a pipeline across the northern shore of the Yukon Territory and the Mackenzie Delta as proposed in the Arctic Gas route.

The NEB stated that "the Dawson diversion is, in the Board's opinion, preferred," and went on to say "the Board would condition a certificate to Foothills [Yukon] to require that the route of the said pipeline within Canada be that route. . . ."[2] The issue of a Dawson diversion may be negotiable, but clearly its implications should be evaluated.

All of the alternative routes are shown in Figure 5. This map also shows the mountain ranges that must be crossed by each route.

The pipeline and tanker mileage of each line from Prudhoe Bay to the lower forty-eight states is shown in Table 2. Of the two all-pipeline routes, Arctic Gas is the shorter by 254 miles; if Alcan were required to reroute its line via Dawson, the difference would become 374 miles. Furthermore, an Arctic Gas line would tap Mackenzie Delta gas and Alcan would not. If a Dempster lateral were added at some later date to transport Mackenzie Delta gas to Canadian markets, the Alcan system would be 834 miles longer than Arctic Gas. An Alcan #2 pipeline in combination with a Maple Leaf line to tap Mackenzie Delta gas would make the Alcan system 1,299 miles longer than an Arctic Gas route

---

[2] National Energy Board, *Reasons for Decision, Northern Pipelines,* vol. 1 (June 1977), p. 1-167.

14

Figure 4

PIPELINE ROUTES FOR ALCAN #2 AND MODIFIED ALCAN #2 VIA DAWSON, WITH A DEMPSTER HIGHWAY LATERAL

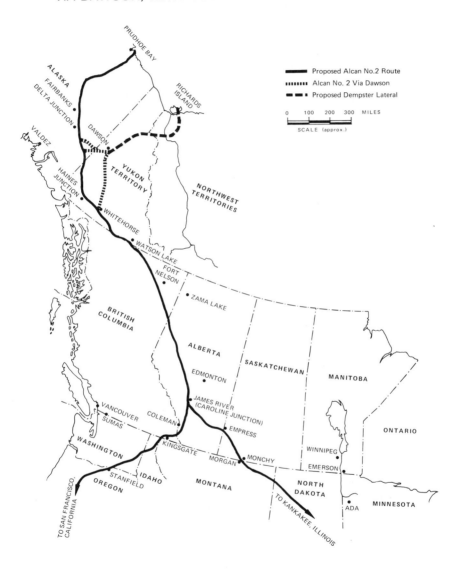

Source: Federal Power Commission, modified by the author.

Figure 5

PROPOSED ARCTIC NATURAL GAS DELIVERY SYSTEMS

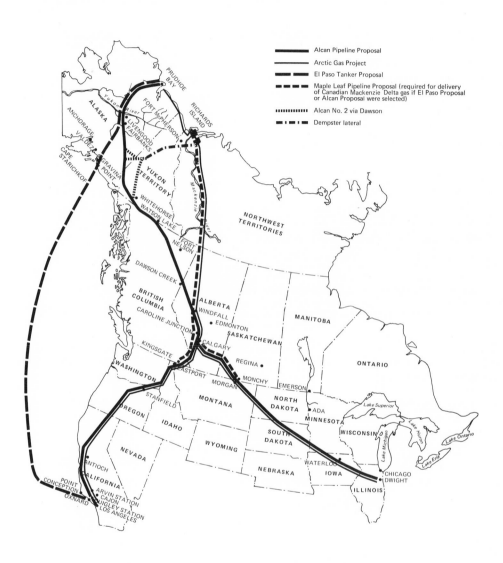

Source: Arctic Gas, *Submission of the Arctic Gas Project Relative to Selection of an Alaskan Natural Gas Transportation System,* July 1, 1977.

Table 2

PIPELINE AND TANKER MILEAGE TO
THE LOWER FORTY-EIGHT STATES

Applicant	Pipeline Mileage			Tanker Mileage (in statute miles)
	Alaska	**Canada**	**Total**	
Arctic Gas	195	2,304[a]	2,499[a]	—
El Paso	809	—	809	2200
Alcan #2	731	2,022[b]	2,753[b]	—
Alcan #2 via Dawson	730	2,143	2,873	—
Alcan #2 via Dawson with Dempster Lateral	730	2,603	3,333	—
Alcan #2 plus Maple Leaf Project	731	3,067	3,798	—

[a] Including a western leg from Caroline Junction, Alberta, to the U.S. border. Excluding the western leg in Canada, the Canadian mileage would be 2,022 and the total Alaskan and Canadian mileage would be 2,217.
[b] Including a western leg in Canada. Without this western leg, the Canadian mileage would be 1,741 and the total Alaskan and Canadian mileage would be 2,472.
Source: Federal Power Commission, *Recommendation to the President*, May 1, 1977, pp. II-3, II-8, and II-14; Alcan Pipeline Company, "Dempster Highway 30" Pipeline Study from the Mackenzie Delta to Connect into a Rerouted 48" Alaska Highway Pipeline near Dawson," undated, p. 3-2.

accomplishing the same result. Finally, the construction of a buried pipeline across discontinuous permafrost areas poses complex "frost heave" problems not yet fully solved, that are likely to be expensive to overcome. All three routes must cross discontinuous permafrost for varying distances, 250 miles for Arctic Gas, 180 miles for Alcan #2, and 100 miles for El Paso.[3] With a Dempster lateral, the Alcan system would be forced to make two crossings of these areas.

Estimates of Gas Shrinkage

The Arctic Gas proposal appears to have a slight advantage over the Alcan #2 proposal with respect to gas shrinkage. Three estimates of

---

[3] White House Task Force, *Construction Delay and Cost Overruns*, July 1, 1977, pp. 46-49.

shrinkage by project are shown in Table 3. The FPC estimates show identical shrinkage percentages for Arctic and Alcan. However, these estimates used a scaling-up procedure based on an Arctic Gas estimate in which throughput volume was 2.25 Bcf/d. In the view of a White House task force report, "This FPC estimate is probably too high."[4] Estimates by applicants and the Department of the Interior indicate that shrinkage in the Alcan project will be from 11 to 15 percent greater than for the Arctic Gas project.

There are three primary reasons for the lower gas shrinkage levels of the Arctic Gas project. First, as indicated earlier, the Arctic Gas project from Prudhoe Bay to the Canadian-U.S. border is shorter; and second, the Arctic Gas pipeline operating pressure is higher (1680 psig, consistent with its .60 inch wall thickness) than that proposed by Alcan (1280 psig, consistent with its .72 inch wall thickness). Also, the Arctic Gas route is both straighter and more level than the Alcan route. In the event that Alcan #2 was rerouted via Dawson, gas shrinkage on this route would be even higher because of the 4.4 percent increase in mileage. Offsetting these Arctic Gas advantages in part is the fact that this project would use more gas in its correction for frost heave than would be required by Alcan (without the Dempster lateral) or El Paso.

The El Paso proposal would suffer from heavy gas usage in the liquefaction plant, which is estimated to consume 5.34 percent of input volume. El Paso's fleet of LNG tankers would consume another 1.59 percent of input volume as LNG boil-off. Its regasification plant would consume 0.08 percent of the gas input volume. In addition, the El Paso LNG plant, ships, and regasification plant would consume fuels other than gas (bunker fuel in ships, diesel in the LNG plant, and electricity in the regasification facility) equal to 1.57 percent of the input volume. While the liquefaction plant and the regasification plant reflect penalties borne by El Paso and not by its two competitive projects, the fuel used in the LNG carriers appears to be approximately equal to the fuel used in the gas pipelines as an alternative means of transportation.

If a Dempster Highway lateral were ultimately appended to an Alcan route via Dawson, shrinkage would increase sharply for Alcan to about 11.3 percent because higher volumes (Mackenzie production plus Prudhoe Bay) would have to be forced through a forty-eight-inch pipeline built with a 1260 psig rating, rather than the 1680 psig proposed by Arctic Gas. If the system from Dawson south were constructed for the higher test strength, its costs would increase. If there were no Mackenzie Delta production for a decade or two, then Ameri-

---

[4] White House Task Force, *National Economic Impact of Alaskan Natural Gas Transportation Systems*, June 30, 1977, p. 14.

Table 3
ESTIMATES OF GAS SHRINKAGE,
ARCTIC GAS, EL PASO, AND ALCAN #2 PROPOSALS
(gas used as fuel stated as percentage of input)

	Source of Shrinkage Estimate				
Applicant	Federal Power Commission	Applicant	Department of the Interior	White House Task Force	Mead
Arctic Gas	6.3[a]	5.5[a]	5.7[b]	5.5[a]	—
El Paso	10.9[c]	10.9[c]	10.9[d]	10.9[c]	—
Alcan #2	6.3[d]	6.4[d]	6.3[d]	6.3[d]	—
Alcan #2 via Dawson	—	—	—	—	6.6[d]
Alcan #2 via Dawson with Dempster Lateral	—	—	—	—	11.3[d]

[a] Prudhoe Bay flow, 2.4 Bcf/d; Mackenzie Delta flow, 1.0 Bcf/d.
[b] Prudhoe Bay flow, 2.4 Bcf/d; Mackenzie Delta flow, 0.5 Bcf/d.
[c] Prudhoe Bay flow, 2.3614 Bcf/d.
[d] Prudhoe Bay flow, 2.4 Bcf/d.

Source: White House Task Force, *National Economic Impact of Alaskan Natural Gas Transportation Systems,* June 30, 1977, p. 15.

can consumers would be forced to pay for capacity that was not utilized.

The different levels of fuel usage, like the mileage differentials between the Arctic Gas and Alcan proposals, are fully reflected in both the cost-of-service estimates and the net-national-economic-benefits estimates to be shown later. The Btu shrinkage data are shown here only as an elaboration of one component of the costs of the alternative systems.

3

THE ENVIRONMENTAL IMPACTS OF THE ALTERNATIVE ROUTES

George W. Rogers

The task of this chapter is complicated by the magnitude of the proposals, their technological differences and complexities, the great geographic distances covered, the diversity of the environmental areas
involved, the size and unevenness of the available data base, and the
lack of consensus among the agencies responsible for protecting the
public interest and the natural environment. The sheer bulk of existing
reports, studies, and supporting documents from which an environmental analysis could be made is impressive, and the task of reviewing
and assessing them sometimes seems greater than the challenge of
compiling them in the first place. The staff of the Federal Power
Commission, for its February 1977 report alone, reviewed 253 volumes
of transcript, embracing almost 45,000 pages and about 1,000 exhibits,
some of which were environmental impact statements of over a
thousand pages.

Assessment of the three competing proposals presents problems
of comparability. For example, is it possible to compare proposals that
employ different technologies (a mix of transport modes or a single
mode), that serve different purposes (transport of Alaskan gas only or
a combination of Alaskan and Canadian gas), and, most important,
that have reached different stages of design development and are
based on uneven research.

In preparing this chapter, reliance had to be placed primarily upon
the draft and final statements of the government agencies charged with
making the official environmental impact assessments. The chapter is
not so much an independent environmental assessment as it is a brief
recapitulation of the main official U.S. and Canadian statements and
the conclusions to which they lead, a review of key elements of the final
environmental impact statements of the Federal Power Commission,
and a critique of the process of assessment.

Principal Assessments

The proposed Arctic Gas and Alcan pipelines would be located for major portions of their length within Canada, and the final official environmental assessments must include the results of investigations by the Canadian government. A preliminary Canadian assessment of the Arctic Gas route was made in 1974 by socioeconomic and environmental teams from the Department of Indian Affairs and Northern Development with a very brief note on the Fairbanks Corridor, that is, roughly the Alcan route.[1] The department was at pains to point out that its work was not strictly an environmental impact assessment. Instead, its purpose was "an appraisal or judgment of the quality of the materials, answers, and proposals submitted by the Application" in the light of environmental aspects in the Expanded Guidelines for Northern Pipelines. The department's conclusion was that the 1974 version of the Arctic Gas proposal was a feasibility report only, "concentrating on principles, theory and assurances," rather than a comprehensive statement supported by complete data and analysis. By November 1976, however, Arctic Gas had submitted a fuller analysis.

The U.S. Department of the Interior through an understanding arrived at in 1974 with the Canadian government produced a final environmental impact statement (FEIS) for the Canadian portions of the Arctic Gas proposal and a briefer statement for the Fairbanks Corridor alternative. This task was somewhat hampered, however, by efforts "to avoid any inadvertent appearance of criticism of the Canadian Government or any implication of gratuitous advice" and by the terms of the agreement, which stipulated "that the statement would be based upon documents and data provided by the Applicant and the Canadian Government, that no studies would be done in Canada by the United States, and that no suggestion would be made for additional studies to be accomplished by the Canadian Government."[2]

The final U.S. environmental impact statements on the proposed Arctic Gas system and the competing El Paso system were the product of a division of labor between agency staffs. The Federal Power Commission staff prepared the final statement for the proposals by the El Paso–Alaska Company and the Western LNG Terminal Company and, "with certain stipulations," used the final statements of the Depart-

---

[1] Pipeline Application Assessment Group, *Mackenzie Valley Pipeline Assessment: Environmental and Socio-Economic Effects of the Proposed Canadian Arctic Gas Pipeline on the Northwest Territories and Yukon* (Ottawa: Ministry of Indian Affairs and Northern Development, November 1974), pp. 153-442.

[2] U.S. Department of the Interior, *Alaska Natural Gas Transportation System: Final Environmental Impact Statement–Canada*, March 1976, pp. 1-2.

ment of the Interior staff as its environmental assessment of the proposal by Arctic Gas.[3]

The Arctic Gas proposal was determined by the FPC in April 1976 to be environmentally preferable to the El Paso proposal, but the FPC's final conclusion was to "*strongly* recommend that *neither* of the applicants' proposals be approved as proposed." (emphasis added)[4] The staff's preferred alternative was the Alcan route without the western leg. Although Alcan had not submitted its application by this date, it was known that the Alcan route was under study. It was proposed that Mackenzie Delta gas, once available, be attached to the system by a 756-mile-long lateral pipeline along the Dempster Highway corridor.[5] The Department of the Interior staff also appeared to favor the Alcan route over the two for which applications had already been submitted, and from its earliest assessments it included comparisons with the Arctic Gas and El Paso proposals almost as though a third application was certain to materialize.[6]

On May 10, 1976, the Northwest Pipeline Corporation did propose a joint United States–Canada undertaking that followed this corridor and used existing Canadian gas systems for final transmission to U.S. markets.[7] The application was filed on July 9, 1976, and the FPC staff issued a supplement to its April 1976 FEIS on September 15, 1976.

On February 1, 1977, the FPC staff published its initial decision. The argument in support of the staff's recommendation was based upon a broadening of the gas reserves to be transported to include not only Prudhoe Bay gas, but production from all probable reserves of the North Slope and Mackenzie Delta. The environmental assessment is summed up as follows (emphasis added):

> The Commission Staff, after (1) supporting its own Fairbanks alternative as environmentally best and (2) asserting that approval of any of the applications will result in substantial environmental impacts, chooses the *Arctic Gas proposal as the least undesirable application*, then Alcan, and the El Paso. Its

---

[3] Federal Power Commission Staff, *Alaska Natural Gas Transportation Systems: Final Environmental Impact Statement*, four volumes, April 1976; U.S. Department of the Interior, *Alaska Natural Gas Transportation Systems: Final Environmental Impact Statement*, nine volumes, March 1976.

[4] Federal Power Commission Staff, *Final Environmental Impact Statement*, pp. I-A7–I-A8. If reasons other than environmental or economic dictate that the route *must* be all-American, the staff still recommends that the El Paso project as proposed *not* be approved.

[5] Ibid, pp. I-A10–I-A11.

[6] See, for example, U.S. Department of the Interior, *Alaska Natural Gas Transportation Systems: A Feasibility Study*, December 1975.

[7] Northwest Pipeline Corporation, "A Statement of Northwest Pipeline Corporation, Transportation of Alaskan Gas, Fairbanks Corridor Route," May 10, 1976.

prime arguments relegate El Paso to third place but still environmentally acceptable (with implementation of appropriate mitigative measures) because of the "environmentally dangerous" LNG facilities in both Alaska and California, unsound pipeline route through the pristine Chugach National Forest, tanker routes through sensitive fisheries, heated effluent discharges in Prince William Sound, and incursion of industry into otherwise non-developed Alaskan areas. In its view, Alcan's inability to carry projected volumes of other Alaskan gas as distinct from the initial Prudhoe volumes without another round of construction is seen to be an overwhelming negative aspect. Since Staff views both El Paso's and Alcan's proposals in the need for an additional pipeline to move Delta gas to Canadian markets, it finds that the total combined proposed construction impacts of both pipeline systems (El Paso–Maple Leaf and Alcan–Maple Leaf) render them less desirable. Furthermore, Staff asserts that neither El Paso nor Alcan has done adequate environmental studies or preparation, and it particularly criticizes El Paso's lack of (1) seismic, (2) LNG site geological, (3) general biological, and (4) treated heated effluent investigation of possible adverse impacts at Gravina Point.

Its support of Arctic Gas can best be described as picking the lesser of the evils. . . . Unless all hydrocarbon exploration and development in the Wildlife Range is prohibited, there will be future development within the range and no reason appears for precluding a gas transmission pipeline as inconsistent with the range. As previously stated, while faulting Arctic Gas for ignoring the probable environmental degradation of the Wildlife Range and its unbroken continuum of Arctic ecosystems, it grudgingly admits that substantial effective work and effort has gone into both the design and operation, that for the most part, it will be successful, and that it should be certificated.[8]

Assessments of all three projects by other parties are included in the Federal Power Commission hearings record and exhibits. The most technically detailed are the series of briefs and reply briefs as each applicant attacked its competitors' projects and defended its own. Most of this has been incorporated in the FPC final assessments. The contributions to the record made by the conservation intervenors and the state of Alaska are not so much true assessments as advocacy tracts

[8] Federal Power Commission, *Initial Decision on Competing Applications for an Alaska Natural Gas Transportation Project*, February 1, 1977, pp. 184-185. The staff appears to believe the Fairbanks alternative with a forty-eight-inch pipe to be "environmentally superior" although its economic viability cannot be determined without further study. (p. 250)

which, judging by the FPC staff's reactions to each, probably did their respective causes (Alcan #1 and El Paso) more harm than good in the final appraisal.[9] Although these assessments have been used in the present review, they contributed no new data or analysis.

Presiding Administrative Law Judge Nahum Litt, in his initial decision of February 1, 1977, agreed with the FPC staff in awarding Arctic Gas first place, but knocked Alcan #1 out of the running entirely.

> If Arctic Gas is unable to accept a certificate, this record supports findings that El Paso's proposal, as required to be modified by the findings above, would also meet the present and future public convenience and necessity. No findings from this record support even the possibility that a grant of authority to Alcan can be made.

The judge deemed the Alcan #1 proposed financing plan inoperable and its design "neither efficient nor economic since the pipeline is undersized."[10]

On March 8, 1977, Alcan withdrew its forty-two-inch proposal and filed a new forty-eight-inch express line proposal to carry only Alaskan gas to the continental United States. The FPC staff and judge's objections to the lack of solid financing and the failure to include transport of Mackenzie Delta gas, however, remained. In a report to President Carter on May 2, 1977, four FPC commissioners unanimously favored a Canadian overland system, but split two to two in the choice between Arctic Gas and Alcan. As to environmental issues they found:

> Each system will have some adverse environmental impacts. We believe all of these impacts to be acceptable, given proper precautionary measures. Arctic would involve crossing the Arctic National Wildlife Range, and other lands now little used by man. The other projects would generally follow existing utility corridors—a distinct environmental advantage.[11]

---

[9] The state of Alaska's main arguments and supporting calculations were summarized in reports of the Office of Pipeline Coordinator and the Gas Pipeline Task Force chaired by the state attorney general: Office of Pipeline Coordinator, *Preliminary Assessment of Three Competing Gas Line Proposals in Alaska*, September 1976; idem, *Comparative Study, Gas Pipeline Proposals of Arctic Gas, El Paso, Northwest*, undated, limited to economic analysis; Gas Pipeline Task Force, *Prudhoe Bay Natural Gas Pipeline Systems*, State of Alaska, Department of Law, November 1976.

[10] Federal Power Commission, *Initial Decision*, p. 430.

[11] Federal Power Commission, *Recommendation to the President: Alaska Natural Gas Transportation System*, May 1, 1977. Quoted material from letter from commissioners to the President, May 2, 1977.

The long-awaited report of the Mackenzie Valley Pipeline Inquiry (the Berger Commission) was released on May 9, 1977, after two years of intensive hearings. It concluded that for environmental reasons no northern gas pipeline should be built and that construction of a pipeline south from the Mackenzie Delta via the Mackenzie Valley should be postponed for ten years to allow for settlement of native claims. Although not endorsing the Alcan route because "an assessment of social and economic impact must still be made, and native claims have not been settled," the Berger Commission concluded that this alternative does not present the same environmental and native claims concerns as the Arctic Gas and Maple Leaf lines because the route has "an established infrastructure" and would not threaten any substantial wildlife species in Alaska or Canada. [12]

The National Energy Board of Canada considered three competing pipeline proposals—the Alcan which would only transport Alaskan gas, the Canadian Arctic Gas Pipeline, Ltd. proposal which would transport Alaskan and Mackenzie Delta gas, and the Foothills Pipe Lines, Ltd. Maple Leaf Project which would transport only Mackenzie Delta gas. In a decision released on July 4, 1977, the NEB rejected the Arctic Gas proposal on grounds that it was environmentally unacceptable, rejected the Maple Leaf Project because of economic and natural gas development and reserves factors, and concluded that Alcan "offers the generally preferred route for moving Alaskan gas." Among other things, the recommended approval of the Canadian portion of the Alcan project was subject to rerouting of the proposed line to pass through Dawson and the undertaking of a feasibility study and filing of an application for a certificate by July 1, 1979, to construct a link along the Dempster Highway between the Mackenzie Delta and Dawson. This rerouting requirement would add about 120 miles to the present Alcan proposal, but would reduce the Mackenzie Delta line by 300 miles. Additional socioeconomic and environmental studies would be required for this Dawson realignment. [13]

On July 1, 1977, the President released the comments of nine federal task forces on the report of the FPC and the three competing proposals. [14] The task force on environmental issues headed by the

---

[12] Mr. Justice Thomas R. Berger, *Northern Frontier, Northern Homeland: The Report of the Mackenzie Valley Pipeline Inquiry* (two volumes) vol. 1 (Ottawa: 1977), pp. xiii-xiv.

[13] National Energy Board, *Reasons for Decision, Northern Pipelines* (three volumes) vol. 1 (Ottawa, June 1977), pp. I-159–I-189. The Dawson alternatives are discussed in detail in chapter 3, vol. 2.

[14] Interagency Task Force on Environmental Issues, *Alaska Natural Gas Transportation System,* Executive Summary, July 1, 1977 and *Report to the President on Environmental Impact of Proposed Alaska Gas Transportation Corridors* by the Council on Environmental Quality, July 1, 1977.

Department of the Interior found that Alcan #2 would have the least environmental impact but also the disadvantage of the longest all-land route and that additional research was needed to define specific problems and appropriate ways of dealing with them. The Council on Environmental Quality (CEQ) agreed with these conclusions, but disagreed strongly with the FPC's conclusion that the Arctic Gas project was environmentally acceptable. This was primarily because it would intrude into a vast wilderness area extending from the Canning River on the North Slope across the Mackenzie Delta and down the Mackenzie Valley and "would be massively disruptive." The CEQ also noted the severe problems of construction and operation in an environment "of unparalleled hostility and fragility" and the effects upon the native peoples of Alaska and Canada. The CEQ gave El Paso low ratings because of risks to public safety, doubts about the integrity of the systems and uncertainties about thermal impacts and seismic design. The Department of Commerce's task force on socioeconomic impacts found that in terms of adverse impacts upon native communities and lifestyles, the Arctic Gas and El Paso proposals suffered in comparison with Alcan #2, but that the El Paso route would probably be most stimulating to the Alaskan economy in general.

Scope of the Present Assessment

In order to bring the assessments within manageable dimensions, this comparison is limited to the three proposals as presented in the FPC staff's decision document of February 1977. Although the Alcan #2 proposal was submitted on March 8, 1977, no assessments of its environmental impact have been made. It will be discussed only in very general terms in the concluding section of this chapter.

The scope of the assessment of each proposal has also been limited geographically to the portions of highest environmental concern. We have considered only those portions north of the approximate points where the transmission systems leave the area of application of the Arctic pipeline construction procedure of the Canadian Expanded Guidelines for Northern Pipelines and enter areas of the Canadian provinces and the lower United States where gas transmission systems are already in existence. For the El Paso proposal, this is the point of regasification in California; for the Alcan proposal, the points at which gas is transferred into the existing systems of the Westcoast Transmission Company, Ltd., and the Alberta Gas Trunk Line Company, Ltd., at Zama Lake, Alberta; and for Arctic Gas, the point where the Canadian Arctic Gas line crosses the Northwest Territories–Alberta border and continues south parallel to the route of the Alberta Gas Trunk Line Company, Ltd. (approximately 60 degrees north latitude).

Whatever project is finally built, additional lines will be required in the lower United States and Canada in order to provide the almost nationwide market spread each seeks. In any case, the addition of new lines in existing corridors would result in relatively minor environmental impact increments. In its 1974 feasibility study the Department of the Interior demonstrated that greater flexibility is available than the applications suggest for the use of displacement and exchange agreements as a means of maximizing the use of existing facilities instead of constructing new ones.[15] Furthermore, for the portions of the projects that have been excluded, the environmental issues that arise are not unique.[16]

An attempt has been made to strike a balance between the *essential* information required for this general assessment and the *exhaustive* information required by the official environmental impact statements. In this matter, the approach used by the 1974 Canadian assessment group was instructive and was approximated in the preliminary review process in identifying environmentally sensitive areas. From a matrix of 19,176 possible interactions between pipeline activities and environmental components, 12 percent were rated as having major, 8 percent medium, 7 percent minor, and 73 percent no potential significance. The topics selected for specific study were all among the 27 percent deemed significant.[17]

The FPC staff's approach to striking a balance between the essential and the exhaustive was more direct and drastic than the matrix approach. The first step was simply to state the key environmental issues as subjectively determined.

> There are three important and controversial "physical" environmental issues in this proceeding affecting Alaska: the Arctic Gas proposal to construct and operate a pipeline across the Arctic National Wildlife Range and El Paso's proposal to both (1) cut across the Chugach National Forest and (2) establish an industrial marine facility in Prince William Sound. On the broader scale, there is only one key "physical" environmental issue: a comparison of the environmental impact of building a project to transport only U.S. gas as against a

---

[15] Department of the Interior, *Alaska Natural Gas Transportation Systems: A Feasibility Study*, pp. 98-127.

[16] The Canadian Pipeline Application Assessment Group, *Mackenzie Valley Pipeline Assessment*, p. 190, cites the rule that Arctic construction techniques are to be used for all work north of 60°N. DOI also adopts this breaking point in much (although not all) of its FEIS analysis.

[17] Pipeline Application Assessment Group, *Mackenzie Valley Pipeline Assessment*, p. 157.

project to transport both U.S. and Canadian gas. The environmental effect of having to build two separate transportation systems, El Paso and Maple Leaf or Alcan–Maple Leaf must be weighed as against the overall environmental impact of building only one.[18]

The remaining assessment task, however, was still awesome. In addition to an abundance of material relating to the physical environment, there were the socioeconomic environmental impacts and the problem of distinguishing between long- and short-term impacts. Some disruptive short-term impacts have no long-term effect, while some minor short-term impacts have totally unacceptable long-term consequences. The final element in the FPC's approach was to assume that maximum mitigation measures would be applied under all three proposals.

> The routes proposed all traverse many areas where there would be only a limited effect upon the environment, either short- or long-term, and no effort will be made to discuss the impact which might occur except in those areas which are particularly sensitive. Similarly, some socio-economic impacts are not so significant as to merit separate discussion. . . . It is also assumed that . . . respective federal, state and local licensing authorities will have a substantial voice as to mitigative environmental activities (such as site-specific avoidance of sensitive areas or timing of construction).
>
> It is further expected that relatively minor realignments or mitigative measures will occur to cure localized events. A discussion of each applicant's route and viable alternative routes from an environmental point of view will be made, therefore, only as to those areas where the environmental effect is significant or the parties have argued it is significant.[19]

The almost crippling differences in the amount of back-up research done by each applicant resulted in further simplifying assumptions. Only Arctic Gas appeared entitled to passing marks for its design and route proposals. The frustration of the FPC staff is tangible throughout its report:

> El Paso, primarily through its lawyers, has made a silk purse out of a sow's ear. It has done little intermediate design work and its design, while impressive on paper and in the strip maps, has no particular backup by core samples or even general site-specific work anywhere along its actual route.

---

[18] Federal Power Commission, *Final Environmental Impact Statement,* p. 175.
[19] Ibid, p. 177.

. . . Alcan itself has not met its burden of proof on construction schedules and its 3-year phased-in construction plan is not supported by the evidence.[20]

In making its assessment, the FPC staff had to fall back on trust— that the competing consortiums were made up of "experienced and knowledgeable companies" and that deficiencies could be made up by "piggybacking" on the evidence of Arctic Gas and the experience of Alyeska.

The three sections following will not attempt any summary of the contents of the environmental material in the record or in the environmental impact statements of the Department of the Interior and the FPC staff, but will sort out and focus on impacts that these sources have identified as significant for the physical, living, and human environment.

Physical Features of the Alternative Systems

Tables 4, 5, and 6 summarize a few of the environmental aspects of the three proposed alternative routes (Alcan #2 is not included)— physiographic regions and divisions crossed by the lines, number of major river crossings, degree of penetration of wilderness, and length of each line through permafrost zones. These are only a sample of the many physical features of the environment which were considered in assessing the safety and potential impact of the engineering design and routing of the three proposals.

The several different physiographic environments through which each line proposes to run, from the Arctic coastal plains to the coastal rain forests in the case of El Paso and to the central plateau country of Alberta in the other two, embrace a rich variety of geological and vegetative zones, each presenting its own special set of engineering challenges and threats of potential environmental degradation if design is not adequate. Problems of soil stability, terrain, and slope characteristics are common to all, but the river crossings and permafrost zones have been chosen here as giving at least an index of the relative difficulties faced by each in terms of physical environmental impacts and design problems.

Major stream crossings are critical points in the lines because of potential unstable soil conditions, disruption of stream beds (including important fish spawning beds), the possibility of introducing chemical substances into the aquatic environment and their spreading through the flow of waters, special construction problems, design problems in increasing the integrity of the pipeline, the correction or avoidance of

---

[20] Ibid, pp. 164-168.

construction problems by stream modification (by culverts, riprapping, channelization, diversion), and so on. Arctic Gas appears to require more stream crossings than the other two lines, but if, as the FPC assumes, both Alcan and El Paso require the construction of a second line (the Maple Leaf Project) in order to move all of the available natural gas, Arctic Gas appears the most desirable in terms of this measure.

Alaskan engineers are familiar with the significant impact of permafrost upon construction—the melting of ground ice, subsidence, thermal and other erosion, destructive drainage changes, topographic and slope fracture, massive soil sloughing, and serious damage to vegetation, to name a few items that come readily to mind. The designers of the Alyeska hot oil line attempted to avoid these undesirable results by elevating much of the line above ground and artificially refrigerating the soil in critical places. All three natural gas pipelines would be basically buried chilled gas lines which should avoid the thaw problems associated with the hot oil line but would introduce an entirely new set of problems. In discontinuous permafrost areas where water migration can occur through the soil to the buried pipe, the below 32 degree Fahrenheit temperature of the gas would cause an ice lens to form, resulting in frost heave, environmental damage, and line instability. The FPC estimates that Arctic Gas would encounter about 250 miles of soils susceptible to frost heave, Alcan 180 miles, and El Paso 100 miles. As in almost every other instance, only Arctic Gas has specifically identified frost heave areas and experimented extensively with means of coping with this problem.

Degradation of the vegetative covering in either continuous or discontinuous permafrost zones will cause thermokarsting (progressive deterioration of the surface until a new equilibrium of heat exchange is established) which would result both in environmental damage and in instability of the buried pipeline. The Arctic Gas line would run for the longest distance through continuous and discontinuous permafrost, with a total of 1,111 miles; Alcan is second with a total of 850 miles, and El Paso third with only 687 miles. But again, if the purpose of the system is to transport all natural gas from the North Slope and the Mackenzie Delta, the Maple Leaf or Dempster Highway lines would be required in conjunction with Alcan and El Paso, and the combination of either set of two lines compared with Arctic Gas gives the single line a distinct advantage. As in the case of the river crossings, this comparison makes clear why the FPC staff believes that there is only one key physical environmental issue: a comparison of the environmental impact of building one consolidated system or two separate systems.

Table 4
ARCTIC GAS ROUTE NORTH OF 60°N, MAIN PHYSICAL ENVIRONMENTAL FEATURES

Physiographic Region or Division	Number of Pipeline Miles	Number of Major River Crossings[a]	Miles through Wilderness[b]	Miles through Permafrost		
				Continuous	Widespread–discontinuous	Fringe–sporadic
ALASKA ARCTIC GAS LINE						
Arctic Coastal Plain	195	17	195	195	—	—
CROSS-DELTA SECTION						
Yukon Coastal Plain						
Prudhoe Supply	130	7	130	130	—	—
Mackenzie Delta						
Richards Island Supply	19	1	19	19	—	—
Prudhoe Supply	47	8	43	47	—	—
Joint Line	89[c]	1	32	87	—	—
Parsons Lake Lateral	15	0	15	15	—	—
Subtotal	300	17	239	300	—	—
REMAINDER TO 60°N						
Anderson Plain	241	4		172	69	—
Mackenzie Plain	359	8	471	—	359	—
Great Slave Plain	160	3		—	80	80
Alberta Plain	58	1		—	—	58
Subtotal	818	16	471	172	508	138
GRAND TOTAL	1,313	50	905	667	508	138

[a] Drainage area greater than 300 square miles.

[b] "Unpopulated and vacant," identified human settlement no closer than 10 miles from right of way.

[c] "Joint Line" continued to approximate boundary between Mackenzie Delta and Anderson Plain.

Source: Applicant's route atlas and data in U.S. Department of Interior, *Alaska Natural Gas Transportation System: Final Environmental Impact Statement*, March 1976, and The Lombard North Group, Ltd., *The Alternatives: A Summary of Three Proposals to Move Prudhoe Bay Gas*, prepared for Alcan Pipeline, August 1976.

Table 5
EL PASO ROUTE, MAIN PHYSICAL ENVIRONMENTAL FEATURES

Physiographic Region or Division	Number of Pipeline Miles	Number of Major River Crossings[a]	Miles through Permafrost		
			Continuous	Widespread– discontinuous	Southern fringe– sporadic
Arctic Coastal Plain	60	0	60	—	—
Arctic Foothills	68	0	68	—	—
Central and Eastern Brooks Range, Ambler Chandler Ridge and Lowlands	122	6	122	—	—
Kokrina-Hodzana Highlands and Rampart Trough	120	2	—	120	—
Yukon-Tanana Upland and Tanana-Kuskokwin Lowland	167	8	—	167	—
North Foothills and Alaska Range	55	8	—	55	—
Gulkana Upland	40	0	—	40	—
Copper River Lowland	55	3	—	55	—
Kenai-Chugach Mountains	123	4	—	—	53
GRAND TOTAL	810	31	250	437	53

[a] Drainage area greater than 300 square miles.

Note: The El Paso route involves approximately 40 miles of line through wilderness, in the Chugach National Forest.

Source: Applicant's route atlas and data in: Federal Power Commission, *Alaska Natural Gas Transportation Systems: Final Environmental Impact Statement*, vols. 1 and 2, April 1, 1976; and The Lombard North Group, Ltd., *The Alternatives: A Summary of Three Proposals*.

Table 6

ALCAN GAS ROUTE, PRUDHOE BAY TO ZAMA LAKE, ALBERTA, MAIN PHYSICAL ENVIRONMENTAL FEATURES

Physiographic Region or Division	Number of Pipeline Miles	Number of Major River Crossings[a]	Miles through Permafrost		
			Continuous	Widespread–discontinuous	Southern fringe–sporadic
ALCAN PIPELINE CO. (Alaska)					
Arctic Coastal Plain	60	0	60	—	—
Arctic Foothills	68	3	68	—	—
Central and Eastern Brooks Range, Ambler Chandler Ridge and Lowland	122	7	122	—	—
Kokrina-Hodzana Highlands and Rampart Trough	120	2	—	120	—
Yukon-Tanana Uplands	135	6	—	135	—
Tanana-Kuskokwin Uplands and Yukon Plateau, North	226	6	—	226	—
Subtotal	731	24	250	481	—
FOOTHILLS PIPE LINES, LTD. (Yukon Territory)					
Tanana-Kuskokwin Upland and Yukon Plateau, North	48	1	—	48	—
Shakwak Valley	142	2	—	71	71
Yukon Plateau, South	221	4	—	—	221
Cassier Mountains	63	0	—	—	63
Laird Plain	35	0	—	—	35
Subtotal	509	7	—	119	390

WEST COAST TRANSMISSION
CO., LTD. (British Columbia)

Laird Plain	92	2	—	—	92
Laird Plateau	57	1	—	—	57
Fort Nelson Lowland	207	5	—	—	207
Subtotal	356	8	—	—	356

ALBERTA GAS TRUNK LINE, LTD.
(Alberta)

Fort Nelson Lowland	50	0	—	—	50
GRAND TOTAL	1,646	39	250	600	796

[a] Drainage area greater than 300 square miles.

Source: Applicant's route atlas and data in: Federal Power Commission, *Alaska Natural Gas Transportation Systems: Final Environmental Impact Statement: Supplement*, September 1976; and The Lombard North Group, Ltd., *The Alternatives: A Summary of Three Proposals.*

One of the major advantages of Arctic Gas is that its route appears to avoid entirely the problems associated with seismic threat. The El Paso line, which traverses several seismically active areas, would face the greatest earthquake rupture hazard. In south central Alaska, El Paso would cross three active linear faults, five alleged inactive faults, and numerous faults in the Chugach Mountains which have not been studied. The final environmental impact statement summary states the matter in a few words:

> In Alaska, the danger of large-scale earthquakes presents serious hazards to the pipeline and LNG plant. Tsunamis resulting from the earthquakes could endanger loading docks and tankers. There is the possibility of the existence of a fault within two miles of the property proposed for LNG facility construction, and this area is on the strike of the major faults involved in the 1964 event. In California, the proposed pipeline route crosses at least 22 mapped fault traces.[21]

In general, the FPC staff found El Paso's seismic research and design to be "preliminary and inadequate" but agrees that this could be overcome. It also noted, however, that with both the oil and natural gas lines traversing approximately the same areas, a seismic event could result in the simultaneous rupturing of both systems for transporting hydrocarbons, while the use of a different route for natural gas would at least insure a continuation of a portion of this flow. The Alcan #2 line would share with El Paso the hazard of proximity to the Denali fault. Design would have to take into account these hazards, but the problem would be minor compared with that faced by El Paso.

The land and native materials (especially gravel) required by the three lines present further measurable physical environmental impacts. In addition to the rights of way for the line itself, a number of support facilities are required in each project which are a function of length, strength, and size of pipe and topography.[22] Within Alaska, Arctic Gas would construct 195 miles of pipeline supported by four compressor stations, two seaports with docking facilities, sixteen aircraft facilities, 250 miles of road, and other miscellaneous sites. The total land requirements would be 4,630 acres of which 3,720 acres would be permanent and used in the operating phase. El Paso's project calls for an 809-mile pipeline with twelve compressor stations and a gas liquefaction facility and tanker terminal. The construction, support

---

[21] Ibid, Appendix F, p. 7.

[22] The discussion in this and the next two paragraphs is based upon ibid., Appendix A, pp. 1-25; and Federal Power Commission, *Alaska Natural Gas Transmission Systems: Final Environmental Impact Statement, Supplement: Alcan Pipeline Project*, September 1976, pp. 12-21.

facilities, and right-of-way land requirements would be 14,712 acres, of which 5,247 acres would be retained for the life of the project. The LNG plant and terminal would require 450 acres. Alcan #2 would construct a total of 731.4 miles of pipeline within Alaska supported by fifteen compressor stations and other facilities (pipe storage, river crossing works, access roads, and so on). The total land requirements are 12,100 acres. Data are not available for the land required on a temporary basis for pipe storage at Prudhoe Bay, Valdez, Fairbanks, and construction camps.

The data on land requirements within Canada are less complete than those for Alaska. The continuation of the Arctic Gas line from the Alaska-Yukon border to the Northwest Territories–Alberta border would be approximately 1,118 miles in length and would require eighteen compressor stations, while the continuation of the Alcan route from the Alaska–Yukon border to Zama Lake, Alberta, would be 914.8 miles in length and would require a total of seventeen to eighteen compressor stations. In California, the El Paso proposal would require 227 acres for its regasification facility, 31 acres for a terminal, 2,250 acres for construction of pipelines from the regasification plant to existing transportation systems (a pair of 142.3-mile-long, forty-two-inch pipelines and one 108.9-mile-long forty-two-inch line) of which 1,300 acres would be permanent. The initial Alcan route included another 817-mile lateral from the Mackenzie Delta to Alberta. This line was rejected by the National Energy Board.

Within Alaska, the estimated Arctic Gas gravel requirements are 3.9 million cubic yards, El Paso 16.4 million cubic yards, and Alcan 11.3 million cubic yards. The continuation of the Arctic Gas line to 60° north in Canada would require approximately 27.6 million cubic yards. Data are not available as to the requirements for Alcan's Canadian extension.[23] The assessments by the Department of the Interior and the applicants appear to indicate that all of the gravel requirements of Arctic Gas and Alcan could be met by use of existing gravel borrow pits created during the construction of the Alyeska oil line and highways near the proposed routes. The FPC staff concluded that "there are localized and even regional shortages" and that "there is in fact a gravel shortage along El Paso's alignment on the North Slope for some 200 miles from the Brooks Range to Prudhoe Bay." To meet this shortage might require stream-bed gravel collection which would severely affect fish and water quality.[24]

[23] Federal Power Commission, *Initial Decision,* Appendix C; Department of the Interior, *Final Environmental Impact Statement—Canada,* p. 353; Federal Power Commission, *Supplement: Alcan Pipeline Project,* p. 9.

[24] Federal Power Commission, *Supplement: Alcan Pipeline Project,* pp. 112-113.

The Arctic Gas engineering design calls for the use of snow as a work pad material in constructing the line and facilities across the North Slope. This technique, though it has been used elsewhere—in particular, on a limited and experimental basis in the constructing of Alyeska—has never been proposed as a major element in a project of this magnitude. If feasible, it has a number of advantages from the environmental standpoint. If the snow pad is adequate, it should impose only temporary and readily corrected damage on the environment and can be made to disappear when no longer needed. The proposal has been received with interest by all of the assessors from the 1974 preliminary Canadian report to the 1977 initial decision of the FPC staff.[25] It has also been a prime target of attack by the other two applicants as well as by the conservation intervenors (who oppose the whole project on principle) and the state of Alaska on the ground of feasibility.

It is proposed by Arctic Gas to collect natural snow by fencing and harvesting from frozen lakes and streams, supplemented by manufactured snow. Snow roads and work pads would be used in all areas of sensitive permafrost north of 65° north latitude, an estimated total of 915 miles. The proposed method of construction is based upon study of past experience in Alaska and the U.S.S.R. and the actual building and testing of three roads at San Sault, Norman Wells, and Inuvik. The FPC staff found that the applicant "has demonstrated, by the weight of vast evidence, that its snow road plan is both feasible and effective and can be accomplished with a minimum of environmental harm." The attractiveness of the snow pad technique is enhanced by the fact "that the alternative is gravel pads, involving problems of borrow sites, aesthetics, permanent accessibility, changes in thermal regime, drainage and added expenses." [26]

The principal environmental impacts of snow roads and work pads are concerned with water withdrawal for snow making and vegetative mat compaction through use of the road. Arctic Gas estimated the need for a total of 8,222,000 bbls. of water for construction from Prudhoe Bay to the west side of the Mackenzie Delta. Of this amount, 6 million bbls. would be for the manufacture of snow. Although the potential sources of water—groundwater, mountain and tundra spring streams, lakes, and ponds—appear adequate, the problem arises from the fact that only spring streams and a few lakes are not frozen to the bed during the winter. These limited sources are also the habitat of spawning and wintering fish. Debate focused, therefore, on

---

[25] Pipeline Application Assessments Group, *Mackenzie Valley Pipeline Assessment*, pp. 233-238; Federal Power Commission, *Initial Decision*, pp. 67-84, 209-223.
[26] Federal Power Commission, *Initial Decision*, pp. 67-68.

the availability of water during the frozen months. Water studies by Arctic Gas have concluded that sufficient water would be available during this critical period without environmental damage if adequate precautions were taken. Water would be withdrawn from lakes which contained no fish and from other lakes according to size, volume of water not frozen, and fish habitat requirements. The FPC staff agreed with this conclusion, though adding that "more precise information concerning water sources must and will be gathered during the final design stages." [27]

The Inuvik tests appear to have demonstrated that a processed snow road could protect the underlying vegetation and that with proper construction secondary impacts of compaction could be avoided or minimized. The Norman Wells test results introduced some ambiguities on this last subject, and it was concluded that although the Inuvik tests were "auspicious," further tests would be desirable. [28]

The El Paso assessment was vague on the safety of the LNG plant at Gravina Point and the marine link with California. Although this proposed system is not based upon any new technology, it involves a significant "scaling-up" of existing technology which would create engineering and operating uncertainties. The most blatant problems are the location of shore facilities in high seismic areas and the introduction of El Paso's own version of super-LNG tankers into waters that already have the heavy traffic of the Alyeska oil tankers. In addition, the tanker route would cross the most active inshore and offshore domestic and foreign commercial fishing areas in the Gulf of Alaska and the main domestic freight lanes from the continental United States into south central Alaska, including an area which has the highest storm frequency in the Northern Hemisphere. The route would skirt the heavily populated area of southern California and the tankers would port at Oxnard terminal, another high seismic area.

With some surprise the FPC staff notes: "No Party has raised the question of LNG technology safety until the closing briefs of the California State Commission and the Conservation Intervenors did so in an almost off-handed manner. Consequently, it was not briefed, although it could have been." El Paso filed a fifty-five-page brief in reply which addressed all LNG safety issues except risk analysis. [29] The second volume of the FEIS by the FPC on the El Paso Alaska System devotes only one-half page to the impact of tanker operations, concerned with the reduction of the area available for commercial fishing through its exclusion from tanker lanes and the destruction of crabbing

[27] Ibid, pp. 208-218.
[28] Ibid, pp. 221-223.
[29] Ibid, p. 93.

gear and salmon seines due to "tanker straying and wave action." The brief Coast Guard risk assessment in an appendix studies the possibility of casualties onshore from an LNG ship accident in Alaska based on tanker safety records in Cook Inlet and Valdez ports for the period 1969-1974. Volume 3 on the California terminal contains a brief accident probability analysis based upon tanker accident statistics for major ports in the United States; it also assesses the probability of a spill following an accident, LNG plume behavior, and the probability of ignition. The probability of fatal accidents (except to members of the crew) is set at or near zero. Although the testimony in the record is highly contradictory, the FPC staff *appears* to conclude that the risk factor is low. The basic problem is that there are as yet insufficient data on LNG handling, and none on LNG handling on the scale proposed, by which the theoretical debate could be resolved.

The FPC staff is more definite about other faults with the proposed system than about the safety risk. The LNG plant would be the largest ever constructed and would involve six processing trains, each 220 percent larger than the largest existing train. The proposed cryogenic tankers would have a 165,000-cubic-meter capacity as compared with the 125,000-cubic-meter capacity of the largest tankers now operating. In the absence of anything beyond "estimates abstracted from an engineering calculation" to back up the applicant's claim that its scaled-up design is feasible, the staff finds that the planned total system of processing, storage, and shipping does not contain sufficient flexibility to ensure reliability of service. "No finding can be made that this proposed design meets the public convenience and necessity. . . . It is simply not credible as presented."[30] In the face of this conclusion, therefore, the matter of safety risk appears academic, which may explain why it was not more thoroughly addressed.

The Living Environment: Wilderness and Wildlife

The record and the final environmental impact statement volumes by the Department of the Interior and the Federal Power Commission devote considerable space to cataloguing environmentally sensitive areas. Where these draw upon the massive research by Arctic Gas they permit a very detailed description of vegetative areas by zone, type, extent, commercial forest harvest potential, wildlife habitat, and so on. It has also been possible to arrive at indices of sensitivity and mitigation recommendations. Similarly, biological species are identified and classified, their relative abundance and habitat requirements quantified and located, their behavior and sensitivity to disturbance noted, and

---

[30] Ibid, pp. 95-99, 137-159. Quoted in conclusion, p. 156.

mitigation measures recommended. Unfortunately, such assessment in depth is possible only for the Arctic Gas case, the other applicants claiming either that their routes avoid the most sensitive areas or that the impact would be minimal or only incremental because existing corridors are used. Rather than attempt a total assessment, therefore, the FPC staff in its initial decision document limited its comparison to wilderness and a few other selected topics.

Of all the environmental concerns expressed, the preservation of wilderness appears to be paramount. If wilderness is defined simply as undeveloped areas, the Arctic Gas route racks up an impressive 905 miles, while the El Paso project comes in second with only 40 miles and Alcan #1 is located entirely within or near existing development. Alcan #2 would involve about 400 miles of virgin country in northern British Columbia and Alberta. The key wilderness concern, however, was narrowed down to two areas—the Arctic National Wildlife Range and undeveloped areas of Chugach National Forest and the proposed Wrangell-Kluane international park complex.

The strongest environmental criticisms of the Arctic Gas proposal in the record relate to its crossing the Arctic National Wildlife Range. The similar intrusion of the El Paso project into a "pristine" area of the Chugach National Forest, on the other hand, received milder criticism. Having made a *cause célebre* of the Wildlife Range, the conservation intervenors, assessing El Paso's entry into the undisturbed wilderness area of the Chugach National Forest, offered only muted opposition and did not infer—as the FPC put it, adopting the conservationists' rhetoric—"that anyone seeking such an invasion is voracious or insensitive to wildlife and wilderness values." The FPC report found that "not one word in the Conservation Intervenors' Brief is even addressed to the environmental consequences of spreading industry through the generally unblemished Alaskan countryside."[31] The explanation, in the FPC staff's opinion, was that the state is "more an advocate on economic grounds than a concerned party merely seeking to protect its environmental heritage."

Granting that wilderness values are in the eye of the beholder, there is some substance to the FPC staff's assessment that "from the point of view of a relatively unimpacted area of both vigorous and spectacular scenic beauty, the mountainous area east of Valdez and Cordova exceeds that of the North Slope."[32] This view is clearly not shared by the state or the conservation intervenors, but it should not be overlooked in making an objective comparison.

---

[31] All quotes from ibid, pp. 191-192.
[32] Ibid, p. 196.

The conservation intervenors' main opposition to the Arctic Gas proposal was that it would destroy one of the nation's finest remaining wildernesses, the Arctic National Wildlife Range. Describing the range in such terms is a totally subjective matter, but even so it is an over-statement, and much of the record was taken up by a detailed rebuttal on the technicality of "wilderness." The FPC staff argued that, quite aside from whether or not it is the finest, it is far from unique.

There are no endangered animals, birds, fish or other fauna or flora in the Wildlife Range, and the same type of animals—whether it be polar bear, caribou or arctic char—can be found throughout the remainder of the 750 miles or so of Arctic Coastal plain. While the ecosystems at any one place are certainly distinctive, peculiar characteristics would be equally noted for each area of the North Slope, just as each human being would "define" a different man. The more subtle appreciations of uniqueness argued here, therefore, are (1) the juxtaposition of certain physical aspects of the coastal plain with the foothills and mountains of the Brooks Range, but for the most part, the same coastal plain, foothills, and mountains occur across the North Slope, and (2) that it, unlike the rest of the North Slope, is "unspoiled."[33]

A review of its history shows that the range is not unimpacted by man and, by definition, cannot be held to be unspoiled. U.S. Navy exploration programs between 1944-1953 and the continuing United States Geological Survey program up until 1964 drew the native popu-lation of the North Slope into Barrow; in the 1939 census Barrow accounted for 23 percent of the region's total native population, in 1970 for 78 percent. Earlier the native peoples had made extensive use of the range area for subsistence, as was demonstrated in hearings support-ing their aboriginal land claims. During the last century, commercial whaling activities along the entire coast, including the wintering over of whaling crews at Herschel Island, drew heavily upon the land as well as the aquatic resources of the region. One consequence was the extermination of the musk ox herds within the range area. Barter Island, Camden Bay, and Demarcation Bay, all within the range, have been used for military purposes, and petroleum exploration both on-shore and offshore is proceeding today.[34] The range has been a popu-lar hunting area and state and native corporations are encouraging guided hunting trips as a means of bringing cash income to the native population.

Compared with man's intrusions, the Arctic Gas pipeline would

---

[33] Ibid.
[34] Ibid, pp. 197-198.

have a relatively low impact in the eyes of the FPC staff. Snow road construction techniques would minimize vegetative damage and reduce the use of gravel from native sites; they would also permit the removal of means of access when construction was completed. Except for four compressor stations, the line would be completely under ground. Clearly the conservation intervenors' charge that a pipeline across the range would "destroy" it is a gross exaggeration.

A further argument advanced by proponents of the Arctic Gas line justifying its route is that the Wildlife Range does not have any legal status under the Wilderness Preservation Act, although its wilderness value was cited in the 1960 public order. Finally, the staff argued that in the context of the national energy and natural gas shortages, "the ultimate incursion into the range for such exploitation must be considered a virtual certainty" and construction of the gas line "becomes a benefit from the point of lessening the environmental cost of attaching the new supply." [35]

Throughout the debate on the preservation of the wilderness, surprisingly little emphasis was placed on the real threat presented to the assets the wilderness areas were created to protect. Opposition to development on the grounds of serving the wilderness is vulnerable, and if it fails, the threatened forms of wildlife must be defended on an individual basis. The range is of greatest importance as the support base for the porcupine caribou herd (about 110,000 to 120,000 animals), which migrates within a 120,000-square-mile area in Canada's Yukon Territory and the Brooks Range and North Slope in Alaska. The most critical areas for the ultimate survival of the herd are the calving areas. The herd calves at Camden Bay on the Beaufort Sea, a barge-unloading and general marshaling area for the Arctic Gas line. The applicant asserts, and the FPC staff apparently agrees, that harm to the caribou could be avoided or mitigated. It is assumed, furthermore, that at all other points on the migration route, the construction and operation of all the proposed systems would not harm caribou herds or could be scheduled to avoid such harm. [36]

Outside the Arctic National Wildlife Range, in Canada, the primary environmental threat posed by the cross-delta route is to snow geese. During a normal year when the North Slope is generally snow-free from late August to late September, 25,000 snow geese stage on the delta, but in a year of very heavy snowfall on the North Slope (averaging about once every eight years), as many as 270,000 geese may arrive at Shallow Bay. There did not appear to be a consensus

---

[35] Ibid, p. 202.
[36] Ibid, p. 205-207.

among the expert witnesses as to whether proposed mitigation measures would provide sufficient protection, and the FPC staff fell back upon the suggestion that if the flock were adversely affected during construction, the hunting of snow geese on the North American continent could be curtailed.[37] Similar concern for the protection of beluga whale-breeding areas on the Delta crossing is not satisfactorily addressed.

The other species that might be adversely affected by pipeline development include polar bear, the barren ground grizzly bear, the dall sheep, migratory birds, and raptors (gyrfalcons, peregrine falcons, eagles, and hawks) in Alaska and Canada. In California, the endangered San Joaquin kit fox might be affected, and the use of vehicles and the operation of the regasification plant might adversely affect the prairie falcon.

The evidence indicates that the Arctic National Wildlife Range is not unique and that a pipeline would not destroy it. What is at stake here is that a pipeline would violate the wilderness principle. But this does not establish damage to the environment.

A final word must be said on the heated water and effluent discharge of the proposed El Paso LNG plant at Gravina Point. As proposed, the plant would have a once-through sea-water cooling system. After intake an algaecide would be added to the water, and before discharge a neutralizing agent would be introduced. The returning water would average 20.7° Fahrenheit warmer than the water in Orca Bay. The minimum design flow would be 490,000 gallons per minute (equivalent to a forty-foot by forty-foot four-story building), and the final design flow would be 658,000 gallons per minute. The record shows that El Paso has done none of the baseline oceanographic population and temperature tolerance studies necessary for an affirmative finding as to acceptable impact upon the marine biota of Prince William Sound.[38]

The Human Environment: Socioeconomic Impacts

The National Environmental Policy Act also requires discussion and analysis of the socioeconomic impacts of major federal actions. The arguments presented in the record concerning the various socioeconomic impacts did not touch on net national benefits, but were limited to Alaska and focused on certain aspects of population, employment, gross state product, personal income, and government revenues and

---

[37] Ibid, pp. 87-90. The Delta appears from the record to be much more sensitive to disruptive impacts than the range, but it has not engendered comparable concern.
[38] Ibid, pp. 240-242.

44

expenditures. The several parties to these arguments used different data and analytical systems and put different weights on the elements of their own findings in arriving at their conclusions. For example, the state of Alaska concludes its employment and income analysis with the following findings supporting the general conclusion that El Paso is best for Alaska:

> To summarize, the El Paso line offers the largest number of employment opportunities (including the long-term jobs which are associated with the LNG facility), and thus, the highest contribution to *total* personal income. The Alcan alternative also supplies high employment and income potentials, although less than does El Paso. The Arctic Gas proposal results in the least change in *total* personal income since it involves far fewer jobs. (emphasis added) [39]

When this comparison is made on the basis of change in real *per capita* income generated by the same model, however, the reverse findings result: Arctic would *increase* real per capita income over the analysis period (1977-1990), while both El Paso and Alcan would result in a *decline* over the period! This is noted in the task force brief but discounted in the discussion immediately following:

> Many of the overall costs and benefits of the gas line choices emerge from the studies undertaken, but there is little solid information concerning who precisely it is that gains and loses from this change in employment. Thus, it cannot be assumed that the increase in per capita income will provide most resident Alaskans with improved financial positions. [40]

Because of "other conditions," the report prefers to measure benefits in total rather than in per capita terms. This is curious. By the same reasoning, it can be argued that *total* personal income increase cannot be assumed to improve the financial lot of resident Alaskans. Obviously, the choice was made on the basis of which set of comparisons appeared to most benefit El Paso.

The FPC staff, in its review, cites further examples of practices which tend to "cook" the results of the analysis, even though they appear to emerge from impartial econometric models: ". . . inputs into the equation were often arbitrary or based on unsupportable assumptions. Even assuming the inputs were not arbitrary, the analysis itself is incomplete (perhaps necessarily), since only direct, foreseeable costs and expenditures were included." The FPC found that in calculating

---

[39] Gas Pipeline Task Force, *Comparative Evaluation, Prudhoe Bay Natural Gas Pipeline Systems,* State of Alaska, Department of Law, November 1976, p. 28.

[40] Ibid, pp. 25-27, 29.

induced employment, the "selection of the multiplier is so arbitrary as to be almost useless." For example, "El Paso uses a multiplier of 35.83 to predict that 21,000 total jobs will be created from about 600 direct jobs in 1983. . . . Arctic Gas suggests that only a 1.5 multiplier should be employed, which is the most common multiplier used in Alaska." [41] When the calculations extend to population, the FPC staff observes: "Merely to say that the projected increases suggested by the parties vary is an understatement. El Paso estimates a population increase from its project of 57,000 by 1980 [the assumed peak year of pipeline construction], while Staff estimates 24,000 and the state of Alaska, 46,000." [42]

The state of Alaska concluded its socioeconomic brief by saying:

> When the benefits of the El Paso project so far outweigh any of its costs, and when its net socio-economic benefits would so far exceed those of the Arctic Gas projects, the State cannot conceive of refusing to grant the El Paso project a preference on socio-economic grounds. [43]

Nowhere in the briefs of the state, El Paso, or the record is there any data or analysis which would support such sweeping conclusions. All that are presented anywhere are mere fragments of the type of analysis required to arrive at such conclusions. The FPC staff is correct in its statement that:

> The record does not permit a 'bottom line' comparison of net benefits to the State from alternative proposals. . . . Conclusions about net benefit rankings should be based on a benefit-cost analysis using methods normally applied by economists in benefit-cost analysis. . . . This, according to the Staff, is missing from this case. [44]

It is indeed missing, and the social cost-benefit analysis for each project is even farther out of reach than the economic.

The FPC staff is probably correct in concluding that all three projects would provide a net benefit to the state. It is also correct in stating that:

> The most substantial economic benefits to the state will come from hydrocarbon severance taxes and royalty payments. It is obvious that these sums are so large as to overwhelm any associated costs of the project once operations

---

[41] Federal Power Commission, *Initial Decision*, pp. 257-258.

[42] Ibid, p. 258.

[43] "Alaska Initial Socio-Economic Brief," quoted in ibid, p. 254.

[44] Ibid, p. 255.

begin. Assuming only 2.25 Bcf/d rate of flow from the Prudhoe field, and a wellhead price for gas of $1.00 per Mcf, Alaska's 12.5 percent royalty interest and 4 percent severance tax will net the state $135.5 million a year. . . . *Once the fact is accepted that all the projects will produce approximately the same revenues in severance taxes and royalties, other economic variables become somewhat less significant.* The bottom line for the state's development is the manner in which the state expends the billions in revenue. It is those state policies which will determine the ultimate socio-economic effects of the projects. (emphasis added)[45]

The record includes discussion and material on the impact of the projects on local communities, particularly those where the population is largely or wholly native (Eskimo, Indian, or Aleut). These are mixed cash-subsistence economies and the threat of any of the projects is stated in terms of depletion or diminution of wildlife for subsistence harvest, increased dependence upon cash, and changes in cultural values and lifestyle. In view of the earlier introduction of the cash economy, the construction and operation of defense facilities, oil and gas exploration and development, the construction of the Alyeska oil pipeline, the Alaska Native Land Claims Settlement Act, the organization and funding of regional and village corporations, and the development of local self-government, these impacts on native communities will be incremental at most. The only difference between the three projects will be variations in the geographic distribution of their impacts.

General Observations on the Process of Environmental Assessment

A professional social scientist can only review with dismay the sections of the FPC's initial decision report dealing with the socioeconomic impacts of the three competing projects. The results are, for the most part, merely projected sets of data on income levels, employment, and government finance to replace or extend the historical data. They do not reflect changes in the hierarchy of systems from which these results emerge. The real difficulty arises, however, when the several socioeconomic assessments are translated into briefs to be used as ammunition in an adversary proceeding. These proceedings seem to be based on the theory that truth will emerge from combat between adversaries on a field of honor, God being on the side of the survivor. What this

---

[45] Ibid, pp. 256-257.

does to otherwise sound professional work is illustrated in some of the passages quoted here.

There is a more fundamental problem that goes beyond the distortion and suppression of evidence common to all adversary proceedings, beyond differences of opinion among expert witnesses and the problem of interpreting findings. It even goes beyond the missing benefit-cost analysis and other methods of analysis normally applied by economists. What is missing is an understanding of the interactions that arise when a man-made physical system, in this case a system for the transport of natural gas, is introduced into a given geographic area where other man-made social and economic systems exist. In some instances the existing social and economic systems are modified, in others they are completely transformed. Whether this is good or bad must be judged according to the value systems held by those affected by this process.

This shortcoming goes beyond the socioeconomic issues and applies to the physical and biological ones as well. In the Department of the Interior and Federal Power Commission's final environmental impact statements (as well as in statements for other projects), a major omission is analysis of the proposals' impact upon existing natural ecosystems. For the most part the underlying analytical framework of these and other environmental impact statements has been implicitly or explicitly some variation of a matrix, a catalogue of impacting and impacted variables. Although using this format for its EIS, the Department of the Interior staff recognized the limitations of the two-dimensional matrix and would agree with the National Research Council's Outer Continental Shelf Impact Review Committee that more appropriate means must be developed to assess the character and productivity of ecosystems and their modification by man's activities. This type of assessment is not now possible. Of the total 778 pages in the Alaska volume of the department's EIS, only 14 pages are devoted to a general recognition of the existence of ecological considerations and only 33 pages of the 825-page Canada volume. The department's staff explained the deficiency:

> The classification of ecosystems, their descriptions, the explanations of their processes, and the synthesis of component parts into total units have never been accomplished for the regions of interest and the intricate relationships forbid hasty interpretations of systems which at best are poorly known.
> . . . The fact is, ecosystems are rather intractable objects of study. . . . The disruption of the functions of a plant community, a lemming population, or a caribou herd are primary effects but the secondary and tertiary effects pulse through

the entire ecosystems along all functional pathways of community inter-relationships.[46]

The space devoted by the department to the impact of the project upon natural ecosystems does not reflect its assessment of their importance.

> There is no possibility of predicting the total impact of the proposed project on any major ecosystems that would be affected. The changes would be generally adverse and perhaps the greatest overall result would be the toll taken in the shrinking number of ecosystems in their natural state. All systems are capable of absorbing some change and recovering, but how much change is not known and it would be prudent to know. In the absence of predictive capacity, it becomes a crucial question whether or not the project would foreclose future options of preserving natural communities at major ecosystem scales.[47]

For the present it appears that our assessments cannot get much beyond "a catalog of environmental parameters."

A review of the record leaves the distinct impression that the best of all possible solutions to the problem of moving our frontier natural gas to market with the least adverse environmental impact has not been reached. Two of the applicants use the corridor concept as though it automatically ended all argument about environmental impact. The FPC staff in its final report appears to throw up its hands before the massive record on environmental impacts. Ultimately it takes refuge in the argument that one system is better than two systems, the necessary sleight of hand being simply to assume that the purpose is to transport both Canadian and U.S. gas. Before this claim, all arguments for the El Paso and the Alcan proposals fall.

It is also unfortunate that the FPC staff concluded at the onset of its environmental assessment that "no useful purpose would be served by any attempt to give a detailed summary of that environmental material contained in the evidence of the parties and in the environmental impact statements of the Department of the Interior (DOI) or the Commission staff."[48] In the decision statement there are hints that the staff was divided on the environmental superiority of Arctic Gas, its final choice. "The DOI and staff environmental witnesses, relying only on environmental considerations and looking solely at Alaska . . .

---

[46] U.S. Department of the Interior, *Final Environmental Impact Statement—Canada*, pp. 243, 246.
[47] Ibid, p. 448.
[48] Ibid, p. 176.

favor a Fairbanks alternative, something akin to Alcan in Alaska."[49]
And at another place,

> In its Position Brief, Staff states that the Fairbanks alternative with the Richards Island Lateral [a 756-mile pipeline from Richards Island to Whitehorse, Yukon Territory] is environmentally superior to the Arctic Gas, Alcan or El Paso systems. . . . The environmental Staff of the FPC concluded (by a 3-to-1 vote) that the Fairbanks alternative with a Richards Island Lateral is superior to the Arctic Gas prime route. . . .[50]

Unfortunately none of this is developed in the report, and the suggestion that the environmental staff or its witnesses might have something important to say is dismissed by reference to the lack of supporting environmental studies, economic feasibility studies (in some places this appears to be based only on inability to do any cost analysis absent a design), and adherence to the requirement that the system also transport Mackenzie Delta gas via a Maple Leaf route. The apparently environmentally superior Dempster Highway lateral had not been proposed by anyone but the DOI and FPC staff and the National Energy Board in July 1977. Therefore, it was not considered in the FPC hearings.

Further, the environmental analysis made no attempt to estimate the economic cost of environmental change. The analytical tools for quantifying the total cost are primitive, of course. However, until the total environmental cost of alternative routes is quantified, we have no basis for judging the tradeoffs. Chapter 5 shows that the Alcan routes reduce net present economic benefits by about $1.5 billion relative to the Arctic Gas route. The critical question becomes the following: are the additional environmental costs of the Arctic Gas route, relative to Alcan, so large that the people should give up $1.5 billion of present value to avoid such environmental costs? In the absence of even a rough estimate of environmental costs, the many volumes of environmental analysis provide no answer to this critical question.

Finally, the reviewer is left with the distinct feeling that other nonconventional transportation modes have not been given adequate consideration. In its initial report the Department of the Interior included analysis of the costs and benefits of alternatives to building any Alaska natural gas transport system (effects of deregulation, development of other natural gas sources, substitute fuels, and so on). In addition to the El Paso alternative (and five variations), which it studied in detail, the department considered other alternative trans-

---

[49] Ibid, p. 246.
[50] Ibid, pp. 250-251.

portation systems (dense-phase pipeline, methanol pipeline, railway, monorail, ice-breaking tanker, submarine, airplane, helifloat, dirigible, and electrical generation at Prudhoe with energy transmission). The Federal Power Commission considered similar alternatives to the two original proposals with more emphasis upon variations of El Paso.[51] Two private proposals were made in 1976 for converting Alaskan natural gas into fuel grade methanol for transport by tanker or in batches through the TAPS line or into methanol with coal to produce a coal slurry for pipeline transport. Neither of these has been the subject of an application to the FPC and neither has been assessed for environmental impact. Now that more is known of the problems and economic costs of the three pipeline proposals, the costs and benefits of these possibly environmentally preferable but nonconventional alternatives could be more attractive than they appeared in 1974 or 1976.

The Final Environmental Assessment

An Associated Press news story dated July 29, 1977, reported:

> The eight U.S. pipeline company members of the Arctic Gas consortium today agreed to support Northwest Pipeline's Alcan Pipeline Project proposal. . . . The Salt Lake-based Northwest Pipeline said today the eight natural gas companies have established a three-member advisory committee to aid Northwest in efforts to obtain government approval of Alcan . . . it was hoped other natural gas companies would join the project to assure broad nationwide representation.[52]

Thus, two proposals appear to have survived the journey from initial application to submission to the President. The El Paso proposal leaves many aspects inadequately evaluated, and Alcan's has been modified in concept by the conditions imposed by the NEB. The Dawson realignment, although adding 120 miles to the total line if routed along the Taylor Highway from Tetlin Junction in Alaska, has a number of environmental advantages. All active seismic faults are avoided, although the route runs parallel to the Tintina fault in the Yukon Territory which carries a much lower risk. The Taylor Highway is being upgraded by Alaska and the additional construction of a pipeline will not disrupt a virgin area unless the Delta Junction cross-

---

[51] U.S. Department of the Interior, *Alaska Natural Gas Transportation System: Final Environmental Impact Statement—Alternatives,* March 1976.
[52] "Alcan Wins Arctic Support," *Southeast Alaska Empire* (Juneau), July 29, 1977.

country alternative is selected. The length of a second Mackenzie Delta pipeline in the future will be reduced by 276 miles and the remainder will be located in a transportation corridor currently under development as a highway. On the minus side there may be adverse impacts upon another caribou herd, and the line would cross a high-use recreation area as well as the proposed Forty Mile Wild and Scenic River System. On balance, however, it would appear that environmentally the Alcan line would be improved.

There is a third alternative that none of the assessments has properly addressed, and that is the measurement of the costs of constructing and operating either of the two proposed systems against leaving the gas in the ground or considering some of the other alternatives listed in the last section of this chapter. This is usually treated as primarily an economic matter, but the environmental and social costs and benefits must be factored into the calculation.

4

THE CANADIAN SOVEREIGNTY ISSUE

Rufus Z. Smith

In the debate over the three proposed routes for a pipeline to bring natural gas from Prudhoe Bay to the lower forty-eight states, much has been made of the issue of Canadian sovereignty. Two of the proposed pipeline routes would cross Canadian territory. For the purposes of this study, these two routes, known as the Arctic Gas and the Alcan proposals, are essentially the same. Their merits need be weighed not against each other, but against the merits of the proposed all-Alaska pipeline, referred to as the El Paso proposal, which would be coupled with transport of liquefied gas by sea.

The arguments fall into two broad categories, one having to do with physical security primarily in the military sense, the other having to do with the broader implications of permitting a link as vital as a natural gas pipeline to be beyond our sovereign control and, instead, subject to the laws and policies of a foreign government.

Physical Security

The proponents of an all-Alaska route have sought to show that it would be less vulnerable to military interdiction and less vulnerable to sabotage than would a Canadian route.

It may be admitted that a shorter overland route, confined to Alaska, would present less of a target to hostile missiles or bombers, presumably from the U.S.S.R., than would a longer route, although it must also be admitted that the target would be closer to the presumed attacker and therefore subject to greater accuracy of fire. Tanker traffic from southern Alaska to the West Coast of the continental United States would also, of course, be vulnerable to enemy attack and would require protection which a totally land-based route would not require.

Expert evaluation of the relative merits of the opposing schemes on these grounds is beyond the competence of this writer, but common sense would seem to indicate that either system would be vulnerable to physical attack in the event of full-scale war, probably nuclear war, and that neither, in such a case, has a clear advantage over the other. At any rate, this is essentially the conclusion arrived at by our own military experts, who, in their portion of the study published by the Department of the Interior and in testimony before the Federal Power Commission, found a near balance in the strategic considerations involved in a comparison of the alternative routes.

Perhaps more to the point, although still highly speculative, are considerations of relative vulnerability to sabotage. It has been argued that an all-Alaska pipeline, being shorter in its land segment and being entirely under United States jurisdiction, would be inherently less vulnerable to sabotage than a route through Canada; such a pipeline would be policed and protected by our own authorities, both state and federal, who might be presumed to be more technically competent and more diligent in the vigorous pursuit of their duties.

History provides little support for the argument that our authorities, either police or military, would be more competent in patrolling and protecting a pipeline across the rugged terrain of Alaska than would Canadian authorities carrying out a similar function with regard to a trans-Canada pipeline. For that matter, while it is of dubious relevance in the last quarter of the twentieth century, the record of success of the Canadian Northwest Mounted Police and its successor organization, the Royal Canadian Mounted Police, in subduing lawlessness in frontier areas throughout the past 150 years, compared with our own record of violence in the period of our westward continental expansion, is such that, if anything, the vote of confidence should go overwhelmingly for a Canadian route.

Nor, in the judgment of this analyst, is there ground for doubting that the Canadian authorities would be adequately motivated to protect a pipeline across Canadian territory. The United States found it entirely satisfactory and in its national interest to rely on Canadian talent and cooperation in this regard during World War II, when we constructed an oil pipeline and a highway, both of high strategic significance, across Canadian territory. We never had reason to regret it. Since the war, Canada and the United States have become formally allied as members of the North Atlantic Treaty Organization and have entered into a cooperative military arrangement, known as the North American Air Defense Command (NORAD), for the air defense of this continent. NORAD, which is alive and well, probably represents a more intimate and successful cooperation between the military forces of two sovereign nations than exists anywhere else in the world. The proposed route of the Arctic Gas pipeline, of course, falls right within NORAD's area of responsibility.

NORAD is supplemented by other U.S.-Canadian politico-military mechanisms of consultation designed to cope with matters of common interest. Notable among them is the Permanent Joint Board on Defense (PJBD), which has been in existence since 1940 and continues to function unostentatiously but effectively to this day, concerning itself with a great variety of matters, including pipelines, related to continental defense. This writer finds no reason to think the United States cannot continue to count on such cooperation.

As to the more general question of Canadian national interest, one should remember that the proposed Arctic Gas pipeline would also carry vital Canadian gas from the Mackenzie Delta to southern Canada. Canada's self-interest in the physical security of the pipeline would be as direct as our own and would not depend on general pledges of mutual defense.

In sum, this writer finds himself in agreement with the Department of Defense, the staff of the Federal Power Commission, and Administrative Law Judge Nahum Litt that there are no persuasive arguments to support the view that strategic military considerations require us in our national interest to confine to U.S. jurisdiction a pipeline bringing natural gas from Prudhoe Bay to the lower forty-eight states.

Political Security

The advocates of the El Paso scheme have argued that it would be a grievous mistake for the United States to embark upon a program for bringing gas southward from Prudhoe Bay which would in any way be subject to Canadian control, on the grounds that Canadians will act in Canadian interest, not American interest.

Canadians will, of course, always act in what they see to be their national interest. The question is whether there is any reason to believe that, with regard to a pipeline from the Alaskan North Slope, their national interest as they see it would be incompatible with ours.

The Canadian government—and here we are talking about any Canadian government, not simply the Trudeau administration—will be obliged to consider carefully a number of complex questions before agreeing to any proposal to join with the United States (meaning both the United States government and a variety of private U.S. corporate entities) in a cooperative venture of the magnitude of the Arctic Gas proposal. Among the considerations are the following: the size of the proposal in terms of economic investment, its implications of boom and bust for Western Canada, the capacity of the Canadian economy to absorb a massive investment over the short term, the consequences for the value of the Canadian dollar and the effect on the competitiveness of Canadian manufactured goods on the world market, environmental and social consequences for the Canadian north and its people, and the political implications of the further heavy involvement of U.S. capital in the Canadian economy. These are questions as profound as any faced by Canada since its decision not to join with us in 1776 in our rebellion against the British. They rival, in their implications, the biggest questions that have confronted the Canadian nation since its inception in 1867.

Current Canadian considerations of these issues stand in notable contrast to the Canadian government's earlier ambiguity and diffidence regarding a trans-Canadian pipeline as an alternative to the Alyeska line to bring American oil from Prudhoe Bay to the lower forty-eight states. In the present instance, there is a clear potential advantage to Canada, in the sense that a Mackenzie Valley pipeline (Arctic Gas) for U.S. gas might also provide the least expensive means of bringing Canadian gas from the Mackenzie Delta to southern Canadian markets.

The attitude of the Canadian government toward a gas pipeline clearly reflects this possibility of direct benefit to Canada. The government, while not prejudging the issue, has long since expressed an interest in fully examining the complex issues; its regulatory and decision-making processes are in full operation; it has negotiated and initialed a pipeline treaty with the United States; it has kept in close consultative touch with the United States government at various levels regarding the matter; and it has synchronized its decision process with that under way in the United States. Neither the United States nor Canada has made its decision, but both clearly regard a trans-Canada pipeline as a possibility, both are keeping their options open, and neither will be taken unawares by the other.

The point of these observations, however, is not, by noting the seriousness and complexity of the issues facing Canada, to cast doubt on the validity of ultimate Canadian cooperation in a joint project. Rather, it is to point out that the implications are being carefully examined in Canada and to assert that if a Canadian government, after full consideration of the issues being debated there, should choose the path of cooperation, then we can accept the offer of partnership with confidence.

Extensive public hearings have been conducted by Justice Thomas Berger heading a royal commission inquiring into the implications of a Mackenzie Valley pipeline for the ecology of the area and for the interests of the native peoples whose lives would be affected. In this connection, some concern has been expressed regarding possible delays in the settlement of claims of the native peoples of the Canadian north and consequent delays in the initiation of construction.

Justice Berger has submitted to the Canadian government the first of two volumes which will constitute his report and recommendations. In it, he declares his unequivocal opposition, on environmental grounds, to any scheme that would bring Prudhoe Bay gas eastward across the Northern Yukon to join with a pipeline to bring U.S. and Canadian gas southward up the Mackenzie Valley. With regard to the Arctic Gas proposal, he recommends a waiting period of ten years in which to examine and settle native claims and to prepare for the social and economic disruptions of pipeline construction. While not endorsing the Alcan proposal, which envisages transport of U.S. gas across the southern Yukon, Berger does not oppose it on environmental grounds and does not appear to have overriding concern about the problem of settling native claims. His second volume, dealing with social and economic implications, is not expected to be completed before the summer of 1977.

It should be borne in mind, of course, that the Berger report has the force only of recommendation and is not binding on the Canadian government. The report will have to be considered by the government and Parliament, and in the end it may very well be that conditions will be imposed which would make either the Arctic Gas or the Alcan proposal uneconomic or would otherwise raise serious doubt as to their feasibility.

In June 1977, the Canadian National Energy Board issued its three-volume report in which it denied the application of the Canadian Arctic Gas Pipelines, Ltd., for a permit to build the section of the proposed Arctic Gas project from the Alaska-Yukon border to the Mackenzie Delta and then south through the Mackenzie Valley to Alberta. The NEB concluded that "the Prime Route of the pipeline along the coast of the northern Yukon is environmentally unacceptable to the Board, as is the Cross-Delta section of that route."[1] Instead, the NEB recommended to the governor in council (that is, to the government) that permits be issued to build a pipeline running from the Mackenzie Delta to the town of Dawson in the Yukon Territory via the Dempster Highway. This route would then connect with a modified Alcan Highway route.

The action of the NEB is not final. The final decision will be made by the government after full debate in the House of Commons and, if the government chooses to modify the conditional approval recom-

---

[1] National Energy Board, *Reasons for Decision*, vol. 1, p. 1-165.

mended by the NEB, after full approval by Parliament as a whole. One may be sure that the views expressed by both the NEB and Justice Berger will not be taken lightly, and one must conclude, therefore, that the probability of Canadian approval is low for the Arctic Gas proposal, and correspondingly high for the modified Alcan proposal.

These are all problems which the Canadian government can be expected to address promptly. It will examine them in the full knowledge that the United States has an alternative in an all-U.S. route and has publicly imposed upon itself a time limit for reaching a decision.

Readers of the transcripts of the hearings before the Federal Power Commission will be aware of the many arguments made by Canadian constitutional lawyers, both pro and con, regarding the power of the federal government of Canada to enter into binding treaty arrangements with the United States which would provide us with the safeguards we need before committing ourselves to a project involving a natural gas lifeline across Canadian territory. Particular attention was devoted to the question of possibly burdensome provincial taxation that might be levied on a Mackenzie Valley pipeline. The Department of State apparently does not regard the problem as of major consequence, since projected provincial and local taxes constitute only a small percentage of the total tax burden. Moreover, the State Department is relying on the pipeline treaty to which both governments have tentatively agreed.

Opponents of a trans-Canada route have argued that under Canadian constitutional practice the federal government cannot prevent irresponsible taxation by provincial authorities. Such taxation, they claim, would be of great ultimate economic harm to the consumers of the gas transported by the pipeline. This argument overlooks the fact that Canadian consumers and other Canadian economic and commercial interests would be associated with the pipeline in a way that would inhibit arbitrary action by the provinces. Moreover, the treaty explicitly contemplates an additional protocol, in the event the two countries approve a trans-Canada route, which would spell out whatever additional safeguards the United States felt were necessary as a precondition to construction.

Adversary arguments were heard by the Federal Power Commission regarding the power of the Canadian government, under its constitutional structure, to enter into or live up to commitments of a kind necessary to protect our interests. No attempt will be made here to summarize the legal arguments put forth, although it should be noted that Administrative Law Judge Litt of the FPC was not persuaded by the gloomy analysis and dire predictions of proponents of the El Paso route.

The real point is that no Canadian government will commit itself to undertakings vis-à-vis our national interests without first assuring itself in both constitutional and practical political terms that it will be able to live up to its commitment. The history of the relationship between the two countries provides abundant proof.

Were these considerations not sufficient, one need only look at the broader nature of the relationship between Canada and the United States. This country has ten times the population size of Canada. Seventy-five percent of Canada's exports are to the United States. Seventy-five percent of its imports come from the United States. One-quarter of all American investment outside the United States is in Canada. Canada has a continuing need for more American investment. We are tied militarily by treaty. Far more important, we are tied by a common heritage and shared values.

It may be argued that in the light of the November 1976 provincial elections in Quebec, Canada faces an uncertain future that might bring political fragmentation; therefore, the United States cannot prudently enter into as vital a cooperative arrangement with a Canadian government as is contemplated in a joint pipeline project. This writer does not accept the view that the political entity we know as Canada is about to crumble. On the other hand, should it crumble, it is difficult to see why we would be unable to protect our interests in Western Canada. Our ability to bargain, to influence, even to intervene, could only be enhanced.

Our present situation is that we have negotiated with Canada a general treaty, yet to be ratified and finally approved by either government, which safeguards reciprocal rights and sets forth mutual responsibilities regarding pipelines through each other's territory. For years, without incident, we have permitted pipelines to carry vital Canadian oil and gas across our territory. Those pipelines remain hostage, if such be needed, to ensure Canadian compliance with the undertakings set forth in the proposed pipeline treaty between our two countries.

Today's world is interdependent. Our economy is linked with those of many nations. We have recognized that our security is linked with that of many others. No nation is linked more closely to us than Canada by history, tradition, geography, economics, and common interest. To argue that we cannot, because of our national security interests, be beholden to a Canadian government to the extent of building a pipeline across Canadian territory is to ignore the realities of the century in which we live.

The question of where to construct a pipeline should turn on other considerations. If the pros and cons of economics and environmental issues indicate that a pipeline should be built across Canada, then by all means let it be built across Canada—provided always, of course, that the Canadian government, in its own sovereign wisdom, chooses to join with us in a cooperative venture on terms we find acceptable and subscribes to the appropriate treaty commitments. If it chooses not to, we have an alternative.

5
ECONOMIC ANALYSIS
OF ALTERNATIVE ROUTES

Procedures Followed to Develop Cost and Benefit Data

There has been no disinterested original analysis of all of the costs and benefits associated with each of the proposed routes. A study of this kind would be enormously expensive. The U.S. Department of the Interior (DOI) contracted with research organizations to analyze the Arctic Gas and El Paso routes and, to a lesser degree, the Alcan #1 alternative. The budget for this analysis was $1.3 million. This budget permitted only a modest reexamination of a few aspects of the alternative routes. Similarly the Federal Power Commission has not had the means to independently assess the cost of the three systems. As a result, the analyses of the FPC and, to a lesser degree, the DOI draw heavily upon cost and benefit data submitted by the applicants.

As a substitute for a thorough independent analysis of the costs and benefits flowing from the alternative routes, the hearing procedure was adopted. Each petitioner presented his estimates of the costs and benefits of his own proposal, and each was then permitted to cross-examine and criticize the claims of others and rebut attacks against his own system. This procedure limits the extent to which any applicant may exaggerate the benefits of his own system. Further, the FPC staff was permitted to cross-examine each witness and to file its own third-party brief. Finally, an administrative law judge representing the FPC cross-examined witnesses, directed that additional information be supplied, and then drew his own conclusions and recommendations. This process is probably an efficient and effective substitute for a more costly independent in-depth study. The Arctic Gas, El Paso, and Alcan #1 proposals, but not Alcan #2, have been examined in this way.

Formal hearings were held continuously from May 1, 1976, through November 13, 1976. A total of 45,000 pages of hearing transcripts and about 100 briefs from the three applicants and the FPC staff, as well as several hundred thousand pages of hearing exhibits, are available to the public and to those who must make the final judgment on the pipeline applications.

Two similar and partially overlapping standard evaluation methods have been used by the DOI and the FPC as well as by the applicants to appraise the costs and benefits of alternative systems.[1] The first approach, strongly favored by the FPC, is to identify the "cost of service." This is basically a regulatory approach. It sets out to identify the tariff or rates that would be charged by the successful applicant and thus emphasizes private costs. The other approach, strongly favored by the DOI, is to assess the net national economic benefits (NNEB) of the alternative proposals. This approach inquires into the social costs and benefits flowing from each project with net benefits discounted to the present in order to obtain a net present value. Our review of the three alternative projects will begin with the cost-of-service approach and will first identify the capital costs of each project.

Capital Cost Estimates

Capital cost estimates are presented for each route in Table 7.[2] These estimates indicate that, first, the capital cost advantage clearly lies with Arctic Gas; second, the Alcan #2 projects have the highest capital cost; and third, the spread from the low-cost project (Arctic Gas with 1 Bcf/d from the Mackenzie Delta) to the high-cost project (Alcan #2 via Dawson) is about $2 billion, or 40 percent.

The Interior Department estimates of capital cost shown in Table 7 assume a combination of 2.4 Bcf/d from Prudhoe Bay and only 0.5 Bcf/d from the Mackenzie Delta. However, the Federal Power Commission concluded that "based upon all of the information available . . . the most realistic level of gas availability from the Mackenzie Delta area that can be projected at this time is 1 Bcf/d."[3] Accordingly an adjustment has been made in the DOI estimates to reflect the higher total cost of this larger Mackenzie Delta volume and the lower U.S. share of this total cost.

A preference for one project over another should not be established on the basis of estimated capital cost. A complete evaluation must rest on both capital and operating costs and consideration must

---

[1] For similar applications, see Charles J. Cicchetti, *Alaskan Oil, Alternative Routes and Markets* (Baltimore: Johns Hopkins University Press, 1972).

[2] All estimates are in terms of mid-1975 costs. Throughout the analysis, future costs and prices are assumed to remain constant, unless otherwise specified.

[3] Federal Power Commission, *Recommendation to the President, Alaska Natural Gas Transportation Systems,* May 1, 1977, p. III-51.

Table 7
DOI ESTIMATES OF CAPITAL COSTS,
ARCTIC GAS, EL PASO, AND ALCAN #2 PROPOSALS
(in millions of 1975 dollars)

| Applicant | Capital Costs | | |
	U.S. share	Canadian share	Total
Arctic Gas			
2.4 Bcf/d Prudhoe Bay plus 0.5 Bcf/d from Mackenzie Delta	5,435	683[a]	6,118[b]
2.4 Bcf/d Prudhoe Bay plus 1.0 Bcf/d from Mackenzie Delta	4,974[a]	1,186[a]	6,160[b]
El Paso			
2.4 Bcf/d Prudhoe Bay	5,622	—	—
Alcan #2			
2.4 Bcf/d Prudhoe Bay	6,419	—	—
Alcan #2 via Dawson			
2.4 Bcf/d Prudhoe Bay	6,963[c]	—	—

[a] Capital costs have been allocated between Canada and the United States on the basis of a standard procedure known as the "Mcf-mile" method. This well-established regulatory procedure allocates to each customer the percentage of total cost of service represented by the volume of gas that each customer is *entitled* to transport, multiplied by the distance in miles by which such gas is transported. This result is then divided by the total volumes which all customers are entitled to transport, and multiplied by the total miles of transportation of all customers.

[b] Using Interior Department cost estimates, costs have been allocated between Canada and the United States on the basis of allocations provided by Arctic Gas. See White House Task Force, *National Economic Impact*, p. 13.

[c] Capital cost estimates for Alcan #2 via Dawson were drawn from Alcan Pipeline Co., *Dempster Highway 30" Pipeline Study from the Mackenzie Delta to Connect into a Rerouted 48" Alaska Highway Pipeline near Dawson*, undated, pp. 3-14.

Note: Capital cost estimates exclude an allowance for funds used during construction and working capital. Neither the Arctic Gas nor the Alcan route includes a western leg in the lower forty-eight states.

Source: The Aerospace Corporation, *Alaskan Natural Gas Transportation Systems, Economic and Risk Analysis*, Supplementary Analysis, June 1977, p. 6.

be given to the timing (present value) of expected capital costs as well as the flow of net benefits.

There are significant differences between the proposed systems. The Arctic Gas and El Paso proposals both call for higher pressure pipelines (1680 and 1670 psig respectively) than the Alcan #2 system (1260 psig). The Arctic Gas and Alcan #2 projects specify a forty-eight-inch-diameter pipeline from Prudhoe Bay, whereas El Paso provides for a forty-two-inch line. The projects differ in routing and in length.

The extent of potential construction-cost overruns is a major issue in both choice of route and in the decision as to whether any gas pipeline from the Arctic ought to be built. Under the adversary proceedings, each applicant must be expected to claim the lowest capital cost he can reasonably defend. Protection for the public rests upon cross-examination by counsel for the competing applicants, questioning by the FPC staff, and possibly the regulatory system which will ultimately be established to govern the rate structure.

The El Paso-Alaska proposal has been criticized intensively by Arctic Gas. The main thrusts of the criticism are that El Paso's seismic designs for the gas pipeline crossing at the Denali Fault and for the LNG plant at Gravina Point are inadequate and that the cost of earthquake safeguard facilities is underestimated. Arctic Gas noted that the LNG plant and marine terminal at Gravina Point would be located in an area where the earthquake potential is extremely high. The 1964 Alaska earthquake was centered in the Prince William Sound area which includes Gravina Point. This earthquake, recording a magnitude of 8.5 on the Richter Scale, caused extensive damage in the Prince William Sound region. The final environmental impact statement prepared by the FPC staff noted that "tectonic displacement of the (LNG) site area as a result of the 1964 event was shown to be substantial. The plant site was raised about 4.5 feet and translated 30 feet to the southeast in response to the regional thrusting. The earthquake epicenter was about 55 miles away."[4]

Given the nature of the seismic problem, Arctic Gas noted that engineering feasibility, safety, reliability, and cost studies were called for, but that "El Paso-Alaska did not make such studies and was content to rely upon superficial assertions concerning seismic design based only on preliminary analysis."[5] In the absence of detailed seismic studies, El Paso-Alaska had indicated that the marine terminal would be founded in bedrock and designed to withstand an earthquake approximating 8.5 on the Richter Scale. The chief criticism voiced by Arctic Gas is that "no seismic design studies were submitted for this record by which the credibility of the El Paso-Alaska assertions could be tested."[6]

Similarly, Arctic Gas was critical of the lack of seismic studies covering the area of the El Paso pipeline, which crosses the Denali Fault

---

[4] Federal Power Commission, *Alaska Natural Gas Transportation Systems, Final Environmental Impact Statement,* vol. II, April 1976, p. 268.

[5] Arctic Gas, "Brief of the Arctic Gas Project Relative to Engineering Considerations," November 12, 1976, pp. 8-9.

[6] Ibid., pp. 9-10.

north of the Alaska Range in central Alaska. An earthquake specialist testifying on behalf of Arctic Gas emphasized the problem of a buried pipeline crossing major earthquake faults. This witness, Dr. Nathan M. Newmark, testified that "to develop a mode or configuration for crossing major faults in underground construction presents a new problem that, heretofore, has not been studied fully for fault motions more than about seven to ten feet, and presents difficulties even in that range."[7] The Denali Fault appears to require a design for up to twenty feet of fault motion.

El Paso-Alaska not only failed to conduct the necessary seismic studies for the southern half of its proposed pipeline and the LNG plant, but also failed to provide for additional construction costs contingent upon earthquake safety requirements. Arctic Gas raised the El Paso capital cost estimate 12 percent to correct for probable overruns.

In addition, Arctic Gas charged that the proposed LNG plant, involving an unprecedented throughput scale as well as complexities of design, would probably lead to delays and therefore increased capital cost. Finally, Arctic Gas charged that the price of the LNG carriers submitted by El Paso was too low and should be raised 9 percent to $220 million per ship.

Upon reviewing the testimony, the FPC staff concluded that "Arctic Gas' contention that El Paso's cost could be at least 12 percent greater appears to be quite reasonable, particularly in view of the greater degree of nontraditional construction involved." While endorsing the Arctic Gas increase in El Paso's capital cost, the FPC noted further that the lack of a proper seismic design data base for the El Paso line "could lead to further overruns."[8]

Arctic Gas and El Paso-Alaska offered similar criticism of the Alcan #1 proposal. In general, the common charge is that the Alcan project southeast of Delta Junction where the Alcan line leaves the Alyeska route is insufficiently defined in critical areas and precludes any credible estimation of the costs of the project. Part of this criticism applies only to the now withdrawn Alcan #1 proposal. However, the Foothills (Yukon) section of the Alcan #2 line running from the Alaska-Yukon border to the British Columbia border parallels and crosses the Shakwak Fault for at least 185 miles. This earthquake fault is apparently not well defined; hence additional research is required before the precise location of the Alcan line in this area can be specified. If the Alcan #2 proposal were modified to pass through Dawson, the

---

[7] Ibid., p. 13.
[8] Federal Power Commission, "Position Brief of the Commission Staff," December 7, 1976, p. 21.

Shakwak fault would be avoided, at a cost to Alcan of another 120 miles of construction ($543 million).

Construction Timing. Timing of the projects differs slightly. All of the proposals assume that a permit will be granted effective January 1, 1978. Arctic Gas and El Paso plan a six-year planning and construction period, beginning with the granting of the permit. This includes a construction period involving three winters and two summers. The Alcan #2 proposal plans for a five-year construction period. The Alcan proposal calls for an initial flow of 1.6 Bcf/d starting on October 1, 1981, followed by a full 2.4 Bcf/d beginning on January 1, 1983. Alcan's three-year schedule prior to its initial gas flow may be unworkable, depending on factors beyond Alcan's control.

Some testimony indicates that the gas-producing firms at Prudhoe Bay may not be able to complete construction of the necessary gathering and processing facilities in time to permit Alcan to deliver gas in its first scheduled year, even if the applicant's own construction schedule is met. This doubt has been confirmed by the FPC staff.[9]

Cost-of-Service Estimates

The cost-of-service concept has evolved out of public utility rate-regulation procedures. It implies standard procedures for depreciating investments over the life of a project and allocating other "necessary" expenses which are considered "acceptable" in rate-making procedures and which must be paid by consumers. The cost-of-service method appears to be favored by the Federal Power Commission in evaluating alternative projects.

Problems of the Cost-of-Service Approach. This approach has several serious flaws. First, it concentrates attention on the costs of the pipeline only and does not include expenses necessary to compensate producers for the gas itself or for such production costs as additional drilling and well installation. Further, the cost-of-service approach does not include the costs of gathering lines and their operation or the costs of processing the gas to make it ready for delivery to one of the proposed pipelines. Because these costs are common to all of the proposals that have been submitted, they are irrelevant to a choice among the routes. The effects of the problems listed below, however, are different for the alternative routes.

---

[9] Ibid., p. 73.

Second, the cost-of-service approach does not consider the timing (present value) of investment outlays or subsequent net income flows. Rather, it seeks to evaluate alternative proposals on the basis of the cost of service for either a single year or an average of several years. This makes evaluation of the alternatives difficult inasmuch as standard cost-of-service procedures cause costs to decline over time as the initial investment is liquidated. The choice of time period used for the evaluation of alternative routes may be largely arbitrary. A present value approach would correct this fault by comparing easily definable net present values.

Third, the cost-of-service approach evaluates only *private* costs incurred by the pipeline operator. For the most part, private costs and social costs correspond one-to-one. However, there are some private costs that are not social costs and should not be considered in evaluating alternative routes. For example, U.S. income taxes which must be paid by the pipeline operator are private costs (transfer payments) but are not social costs. Of course, any governmental expenditures that are incurred purely as a result of the pipeline construction and operation are social costs. Further, from a national rather than a North American point of view, taxes paid indirectly by U.S. consumers of gas to the Canadian government in the form of income or property taxes are real costs to the United States. This problem creates a difficult social cost allocation problem for the three alternative proposals. The Arctic Gas and Alcan proposals require relatively heavy payments of taxes indirectly by U.S. consumers to the Canadian government. The El Paso line is an all-U.S. route and, therefore, involves no payments of taxes to a foreign government.

The opposite relationship also exists: there are situations where some social costs (benefits) are not private costs (benefits). For example, El Paso contemplates receiving a subsidy in the form of U.S. government guarantees of LNG tanker investments under Title XI of the Merchant Marine Act of 1931. This subsidy artificially lowers the cost-of-service calculation for El Paso. Therefore, in a net national economic benefit analysis this subsidy should be counted as a social cost.

Further, the cost of service may not include environmental damage. It does include costs incurred by pipeline companies attempting to avoid or reduce such damage. In an NNEB analysis, the cost of environmental damage should be included. But in many cases, quantification of such costs is not feasible. In this event, environmental damage should be noted qualitatively and set aside to be considered along with the social costs and benefits that are quantified.

There is also a potential social benefit from the delivery of Prudhoe

Bay gas, but a benefit that is not paid for by gas consumers or collected by the pipeline owners. If gas becomes available from the Alaskan Arctic, it becomes part of the "reliable" U.S. energy supply, displacing crude oil imports and reducing the need for a ready reserve of crude oil. It lowers the social cost of the Strategic Petroleum Reserve System established by Congress in 1975. This benefit should be included among the social benefits, but it would not be included in a cost-of-service analysis.

Fourth, the cost-of-service approach depreciates fixed assets over a period of years. This reduces the capital investment and correspondingly reduces the cost of capital that is included in the cost of service. This procedure shows a high cost of service in the early years, declining persistently over the life of the project. As a result, the cost of service will differ widely depending on what year or group of years is quoted. As a solution to this problem, a "levelizing" procedure may be used, in which case the cost of capital is spread evenly over the life of the project. This makes it easier to compare the proposals, and also to judge whether any pipeline should be built.

Cost-of-Service Estimates by the FPC and the DOI. The cost-of-service estimates are intended to show transportation charges that the ultimate consumers will be asked to bear for the transportation of gas from Prudhoe Bay to an average location in the lower forty-eight states. These pipeline transportation charges do not include the wellhead cost of gas, gathering and processing costs at Prudhoe Bay, or retail distribution charges. Further, they are based on cost conditions prevailing in mid-1975 and, therefore, do not include any cost inflation from 1975 to any future delivery date. In spite of these shortcomings, cost-of-service estimates are useful in rating the alternative pipeline proposals for efficiency.

Table 8 provides cost-of-service estimates based on data submitted to the FPC by the applicants. These estimates were developed by the commission in a last minute effort to obtain cost-of-service data based on standardized accounting procedures. Nevertheless, the data published by the commission still reflect substantial differences in gas input between the three projects, as well as some rather minor technical differences. Estimates published by the FPC have been adjusted in Table 8 to reflect a standardized input, except for an insignificant and uncorrectable input problem in the El Paso system. The FPC cost-of-service data for Arctic Gas have been lowered to reflect an increase in pipeline utilization from 2.25 to 2.4 Bcf/d.

Based on cost-of-service data for the first five years of service, Table 8 shows that the Arctic Gas system could transport gas to

Table 8

COST-OF-SERVICE ESTIMATES,
ARCTIC GAS, EL PASO, AND ALCAN #2 PROPOSALS

(in 1975 dollars per million Btu)

Applicant	Average, First Full Five Years	Average, First Full Twenty Years
Arctic Gas	1.12	0.72
El Paso	1.53	1.09
Alcan #2	1.24	0.79
Alcan #2 via Dawson	1.30	0.83

Note: Volume assumptions are as follows: (1) Arctic Gas—Prudhoe Bay, 2.4 Bcf/d, Mackenzie Delta, 1.0 Bcf/d; (2) El Paso—Prudhoe Bay, 2.3614 Bcf/d; (3) Alcan—Prudhoe Bay, 2.4 Bcf/d.

Source: Federal Power Commission, *Recommendation to the President,* May 1, 1977, p. IV-14. These estimates were collected by the FPC from the applicants in response to an FPC request for cost-of-service data based on "identical" assumptions. The data as reported by the FPC were based on an Arctic Gas input of 2.25 Bcf/d from Prudhoe Bay, whereas Alcan input was 2.4 Bcf/d. In order to attain a higher degree of comparability, Arctic Gas estimates were adjusted to reflect 2.4 Bcf/d from Prudhoe Bay. The adjustment was taken from Arctic Gas, *Submission of the Arctic Gas Project Relative to Selection of an Alaskan Natural Gas Transportation System,* July 1, 1977, p. 42. The small difference in El Paso input is accounted for by indivisibilities in the LNG plant. In order to raise the El Paso input to 2.4 Bcf/d, an additional LNG train would be needed, which would be operated at an inefficient capacity. The Alcan #2 via Dawson estimate is based on the Alcan submission, adjusted on the basis of incremental mileage.

wholesale markets in the lower forty-eight states at an average cost of $1.12/million Btu. This is $0.12 (10 percent) less than the next best system, Alcan #2, and $0.41 (27 percent) less than El Paso. If Canada requires that Alcan #2 be modified to run via Dawson in the Yukon Territory as indicated in the National Energy Board report of June 1977, then an additional $542.7 million in capital cost will be required for the Alcan proposal, as well as additional operating cost. The line would be 120 miles longer than estimated in the Alcan #2 system.[10] The cost of service in the early years would be increased $0.06/Mcf.[11] If Alcan #2 were detoured through Dawson, any subsequent Dempster lateral would require construction to Dawson rather than to Whitehorse, a saving of 276.5 miles. The cost-of-service savings to Canadian consumers of Mackenzie Delta gas would be $0.12/Mcf.[12]

---

[10] Alcan Pipeline Company, "Dempster Highway 30″ Pipeline Study from the Mackenzie Delta to Connect into a Rerouted 48″ Alaska Highway Pipeline near Dawson," undated, p. 3-2.

[11] National Energy Board, *Reasons for Decisions,* vol. 1, p. I-167.

[12] Ibid.

The standard practice followed in calculating the cost of service requires that interest charges on capital employed in the project be computed on depreciated investments in plant and equipment. Therefore, the estimated cost of service declines over the life of the project. Accordingly, the cost of service will differ depending on what year or group of years happens to be used as a base. Early-year costs are high, and late-year costs are low. In order to provide a single figure reflecting costs for the minimum life of the project, the average cost of service for a full twenty years for each project is also shown in Table 8. According to these FPC data, the Arctic Gas project would deliver gas at the lowest pipeline cost, both for the first five years and for a twenty-year period. But the Alcan projects are close seconds, with El Paso at a substantial disadvantage.

Interior Department cost-of-service estimates based on the same basic data have introduced a useful innovation. The DOI calculations have "annualized" the data to correct for the depreciation and interest cost problem discussed above. The annualized estimates express cost of service as a single figure providing the same yield on capital as given in the standard practice. The results of the DOI calculations are shown in Table 9. This annualizing, along with other adjustments made by DOI, reduces the Arctic Gas cost advantage over El Paso to 22 percent and almost eliminates the advantage of the Alcan projects over El Paso. This would be particularly true if Alcan were required to build the Dawson diversion.

There are significant differences between the FPC and DOI cost-of-service estimates shown in Tables 8 and 9. The FPC adopted cost estimates submitted by the applicants, whereas the DOI developed its own estimates on the basis of other data submitted by the applicants. A major factor accounting for the differences between the estimates is an adjustment that DOI made, raising Alcan's construction costs for its mainline through Canada. Both independent DOI consultants and extensive testimony indicated that the estimates submitted by Alcan had been far too low. The DOI added $1,516 million to Alcan's capital cost estimates. Further, the applicants had used differing financial plans involving varying degrees of leverage, whereas DOI imposed a uniform financial plan on all three proposals. In sum, the DOI estimates appear to be significantly better than those of the FPC and will be relied on for the remainder of this analysis.

Some contingent charges may be levied against Alcan #2 via Dawson, and perhaps Alcan #2 as well, as a result of levies by the Canadian government. For example, the National Energy Board recommended to the governor in council that a one-time charge not to exceed $200 million be levied against Foothills (Yukon), the Canadian

Table 9

INTERIOR DEPARTMENT ESTIMATES OF COST OF SERVICE,
ARCTIC GAS, EL PASO, AND ALCAN #2 PROPOSALS

(in 1975 dollars per million Btu)

Applicant	Average, Full Twenty Years	Levelized Twenty-Year Average
Arctic Gas	0.70	0.85
El Paso	0.92	1.09
Alcan #2	0.84	1.03
Alcan #2 via Dawson	0.88[a]	1.07[a]

[a] Based on National Energy Board estimate of additional cost due to a Dawson detour. See *Reasons for Decision, Northern Pipelines*, vol. I, p. I-167.

Source: Estimates prepared by the consultant to the Department of the Interior. See The Aerospace Corporation, *Alaskan Natural Gas Transportation Systems, Economic and Risk Analysis*, Supplementary Analysis, June 1977, p. 6.

line scheduled to carry Prudhoe Bay gas from the Alaska-Yukon border through the Yukon Territory to the British Columbia border. This proposed charge is intended to "pay for the socio-economic indirect costs of the pipeline project in the area north of the 60th parallel." The government would use the funds "toward payment of social and economic costs generally attributable to the pipeline project."[13]

On August 2, 1977, another Canadian government commissioned study recommended a $200 million pipeline charge. A three-member panel headed by Kenneth Lysyk, dean of the law faculty at the University of British Columbia, recommended that the pipeline through the Yukon Territory be required to pay $200 million into a "Yukon Heritage Fund" to compensate for the overall impact of the pipeline in the Yukon. In addition, the Lysyk report recommended that construction of the Yukon sector of the Alcan project be delayed two years until August 1, 1981, to permit settlement and partial implementation of plans for Yukon native land claims. This latter recommendation would delay delivery of gas by Alcan until about August 1983, instead of October 1, 1981, as planned by Alcan.

In the event that a full $200 million charge were levied against the Foothills (Yukon) project as part of the Alcan system, the cost expressed as a component of the cost of service would amount to an additional $0.03/Mcf. Because this social and economic charge is only a contingent liability at the moment, it has not been included in the data shown in Table 9.

[13] Ibid., p. I-176.

Delivered Cost of Prudhoe Bay Gas. The cost-of-service estimates are for pipeline transportation only, from wellhead to distributor. These cost estimates do not include payments to producers for the gas they have discovered, the additional cost of developing the Prudhoe Bay field for gas production, or the cost of gathering and processing the gas. These costs must be paid by consumers and added to the cost-of-service estimates as shown above in order to obtain a cost of gas delivered to retail distributors.

Field development costs include the incremental cost of drilling and equipping wells to produce gas, the cost of gathering lines and facilities, and the cost of processing and otherwise preparing gas for delivery to a pipeline. There are also miscellaneous costs such as those connected with water treatment facilities.

In addition, the cost of producing gas from Prudhoe Bay includes a loss from the forgone oil production as the extraction of natural gas reduces pressure in the oil fields. This lost oil output is estimated at 389 million barrels over the life of the field, which, at $9 per barrel at the wellhead, would have a value of $3,501 million.[14]

The sum of all forgone oil values, field development, production, and gathering and processing costs for gas over the life of the field is estimated to be $8,630 million. This has been calculated on a unit cost basis at $0.47 per Mcf.[15]

In addition to all of these real costs, which must be paid by the consumers, we must add a transfer payment (economic rent) in the form of a residual wellhead payment to the producers. In the absence of continued price control by the Federal Power Commission, the residual wellhead price would be determined by subtracting all costs of producing and delivering gas to markets from market determined prices. With continued price controls, the wellhead price will be whatever the FPC determines.

For the purpose of deciding which alternative route is least costly, it makes no difference what wellhead price is used, providing that the same price is used for all routes. For the purpose of determining whether any pipeline should be built, the relevant wellhead price becomes that minimum price at which producers will produce and sell their gas, rather than forgo gas production for sale and reinject the gas into the reservoir. We will arbitrarily set that minimum at $0.05/Mcf.

---

[14] The Federal Power Commission concluded that no oil production would be forgone in the process of producing gas. This conclusion, however, appears to be a mere legalism. The proof cited by the commission is that "the producers' operating agreement indicates that no oil production will be forgone." Federal Power Commission, *Recommendation to the President*, May 1, 1977, p. IV-3.

[15] Testimony by FPC witness J. H. Goldstein, JHG-1, Schedule 6.

The sum of real costs ($0.47/Mcf) and the minimum wellhead price ($0.05/Mcf) amounting to $0.52/Mcf ($0.46/million Btu) can be added to the cost of service in order to estimate the minimum (except for any cost overruns) delivered price to distributors in the lower forty-eight states. The DOI estimates of the levelized twenty-year average cost of service have been used in this calculation as shown in Table 10. The relative positions of the three routes are, of course, unchanged.

The delivered price estimates indicate the lowest possible price at which the producers and each applicant would be able and willing to deliver Prudhoe Bay gas to lower forty-eight state markets. These calculations assume no cost overruns, an issue to be taken up later. Using DOI cost-of-service estimates, both El Paso and Alcan via Dawson could deliver and sell gas at a minimum price of about $1.75/Mcf.

Before we can determine whether any of the proposed lines can deliver Prudhoe Bay gas to lower forty-eight markets at competitive prices, we need natural gas price estimates. Since costs are based on mid-1975 conditions, market prices must be estimated on the same basis.

After a sophisticated analysis which included estimation of demand elasticities as well as supplies from various alternative sources, the DOI concluded that, based on a $12/barrel price of crude oil, the price of Alaskan gas delivered in the lower forty-eight states would rise from about $2.53/Mcf during the early 1980s to around $2.70/Mcf by the year 2000. The Federal Power Commission assumed a constant value of $2.62/Mcf.

On the basis of these mid-1975 estimated gas prices and costs, the analysis indicates that, without cost overruns, any of the three alternative proposals could operate at a profit.

Effects of Inflation on Costs and Prices since mid-1975. Because all construction costs, operating costs, and natural gas prices used in the FPC hearings and in all cost-of-service and net national economic benefit analyses were based on mid-1975 prices, a question arises as to the effect of inflation from 1975 to 1977, as well as the effect of future inflation, on costs and revenues after construction is started.

Construction costs under Arctic conditions increased from mid-1975 through the first quarter of 1977 by approximately 11 percent,[16] and through June 1977 by approximately 14 percent in total. The behavior of natural gas prices in this same period of time is largely irrelevant because interstate gas prices are controlled by the FPC, and intrastate gas prices are influenced by FPC controls. This influence has

---

[16] White House Task Force, *Construction Delay*, p. 32.

Table 10

MINIMUM DELIVERED COST OF GAS TO RETAIL
DISTRIBUTORS IN THE LOWER FORTY-EIGHT STATES

Applicant	Cost of Service, Levelized Twenty-Year Average (per million Btu)	Real Cost[a] (per million Btu)	Minimum Delivered Cost of Gas (average lower forty-eight-state market)	
			Per million Btu	Per Mcf
Arctic Gas	$0.85	$0.46	$1.31	$1.49
El Paso	1.09	0.46	1.55	1.77
Alcan #2	1.03	0.46	1.49	1.70
Alcan #2 via Dawson	1.07	0.46	1.53	1.74

[a] Includes production, gathering, processing, forgone oil, and minimum payments for gas.
Source: Author's calculations. See text, p. 73.

been the result of gas supply diversions from interstate to intrastate markets. Accordingly, one cannot judge gas price movements from the published record. As indicated above, both the DOI and the FPC estimated the mid-1975 gas price by inference from the price of oil at $12/barrel. The OPEC cartel increased oil prices 10 percent effective October 1, 1975, and another 10 percent between January and June 1977, a total of 21 percent from mid-1975 through June 1977.

The conclusion indicated by this analysis is that whether prices and costs are estimated in mid-1975 or first quarter 1977 terms makes little difference. Price movements in the two areas have been approximately the same, with a probable edge for the value of gas.

As construction gets underway, each year a smaller share of the total construction costs will be affected by new inflation and a larger share will become a fixed cost not subject to inflation. Yet the market price of natural gas is likely to continue to increase. When construction is completed, the total plant investment will become a fixed cost. The only costs subject to future inflation will be maintenance, equipment replacement, and operating costs. These costs will be small relative to the value of the natural gas delivered to market, which may be assumed to continue to increase. Therefore, future inflation will probably raise the value of the product more than the cost of delivering it to market.

Summary of the Cost-of-Service Analysis. On the basis of cost of service to consumers in the lower forty-eight states, the Arctic Gas proposal is clearly superior. This is true whether FPC or the more accurate DOI estimates are used, and whether costs are expressed as a first-five-year average, the average of a full twenty years, or levelized costs. The Alcan proposal comes in second, and El Paso third. The National Energy Board may require a Dawson diversion for the planned Alcan #2, reducing its attractiveness from the U.S. consumer's point of view. With a Dawson diversion, the DOI estimates indicate that Alcan and El Paso costs of service would be approximately the same. Any of the alternative routes is capable, in the absence of large cost overruns, of delivering Prudhoe Bay gas to lower forty-eight state markets at costs which are substantially less than the value of gas in those markets.

Net National Economic Benefits Estimates

The preceding analysis in terms of cost of service has been shown because it has a common sense appeal and, in addition, it is the customary way that the Federal Power Commission evaluates pipeline proposals and regulates rates. However, the method cannot truly determine either (1) whether any pipeline should be built from the Arctic, or (2) which alternative pipeline proposal is most beneficial to society.

The alternative transportation system proposals have also been evaluated in terms of their net national economic benefits (NNEB). This approach asks what is the value of the social benefits expected to flow to the nation and what is the value of the social costs to the nation as a result of the alternative proposals? Unlike the cost-of-service approach where an average cost is calculated, the NNEB approach considers social costs and benefits flowing over a period of time and seeks to determine the net present value of future benefits and costs.

Net National Economic Benefits of the Alternative Projects. Estimates of NNEB were prepared by Arctic Gas and El Paso, but not by Alcan, for their respective projects. Based largely on data submitted by the applicants regarding costs and revenues, NNEB estimates have been prepared by the DOI, the FPC, and most recently by the White House task force (WHTF). The WHTF used the FPC methodology, but modified the evaluation in several important respects. The NNEB analysis presented here will draw on the FPC and WHTF analyses, making some residual corrections, to arrive at a composite NNEB statement for each project, including Alcan #2 with a Dawson diversion. The FPC estimates will serve as the base for our estimates. Following established procedures, all costs and benefits will be expressed in terms of mid-1975 values, prices, and costs. A 10 percent discount rate will be used throughout the analysis.

The value of delivered gas was incorrectly estimated by the FPC, which used shrinkage rates of 6.31, 10.9, and 6.32 percent for Arctic Gas, El Paso, and Alcan respectively. This error arose from the FPC's failure to adjust fuel usage as part of the increase in Arctic Gas input from 2.25 to 2.4 Bcf/d. The correction was made in the WHTF report reflecting a 5.51 percent shrinkage for Arctic Gas. This adjustment raised the present value of Arctic Gas benefits by $107 million.

Gas usage for the Alcan #2 via Dawson route will be increased in proportion to the added mileage. The additional 120 miles increase shrinkage to 6.6 percent.

Further, the value of delivered gas for Arctic Gas and Alcan was incorrectly represented as a result of an error in the Btu conversion factor. Because of the higher operating pressure capability of the Arctic Gas pipeline (1680 psig) relative to Alcan (1260), Arctic Gas would be able to transport a higher Btu-rated gas. In the Alcan submission to the FPC, the applicant changed assumptions about the extent of gas processing at the gas-conditioning plant prior to the delivery of the gas into the pipeline. No parallel assumption was made in the Arctic Gas submission. The assumptions should be identical. Accordingly, we have raised the Btu conversion factor for Arctic Gas to 1168 Btu/cf, corresponding to 1138 for Alcan. This leads to a net increase of $221 million present value for the Arctic Gas system. We have shown this as an increase in the value of the gas. The cost of gathering and conditioning for Arctic Gas must be raised by $96 million (present value) as a result of additional conditioning facilities.

The value of delivered gas is also distorted by the early delivery date assumed for the Alcan system (or the assumption of constant natural gas prices through the year 2006). The FPC program adopted a delivery schedule in which gas would begin flowing to the lower forty-eight states on July 1, 1982, and would move to a full 2.4 Bcf/d effective January 1, 1983. If the two-year construction delay for Alcan's Yukon section as recommended by the Lysyk report is accepted, then initial gas deliveries by Alcan would begin about August 1983. In contrast, both the Arctic Gas and El Paso systems are assumed to begin delivery of 2.4 Bcf/d on July 1, 1983.

The record is replete with evidence that the Alcan #1 proposal, submitted very late, was less adequately researched than the other two systems. The modified Alcan #2 project was not introduced until March 8, 1977. It involves not only an enlarged-diameter line, but also

76

about 400 miles of new route in Alberta and the possibility that a Canadian permit might require the Dawson diversion and even more research. Even if permits from Canada and the United States were issued and effective by January 1, 1978, Alcan would have more research and planning to do than would be required for the other lines. Further, as indicated earlier, there are real doubts whether the necessary gathering and processing facilities would be constructed and ready to deliver gas by the producing companies early enough to permit deliveries to any pipeline as early as July 1, 1982. On the other hand, a shorter period of actual construction would be expected for Alcan because, through Alaska, this system would utilize data already developed for Alaska, and through all but about 400 miles of its length a nearby road system already exists.

This matter of timing would not be a significant issue except for the assumption in all studies that the real value of natural gas will remain essentially constant through the year 2006. Assigned a 1975 price of $2.62/Mcf, natural gas starts out being underpriced relative to its substitutes. Moreover, natural gas is a nonrenewable resource competing with other nonrenewable fossil fuels and electricity. Therefore, it is quite likely that its real social value will increase into the foreseeable future at not less than the real interest rate—about 3 percent per year. In this event, it would make very little difference whether one pipeline were scheduled to begin operating sooner than another and consequently depleted its supply sooner. Because of the constant price assumption, it seems wise to assume that all three pipelines could commence deliveries on July 1, 1983. Accordingly, in this analysis, Alcan production, scheduled in the FPC model to begin on July 1, 1982, will be delayed one year and shifted to a common terminal year, 2007. This reduces the present value of Alcan #2 gas value by $1,009 million, making the gas value for 2.4 Bcf/d input equal, except for differences accounted for by shrinkage and Btu rating, in the three lines.

If we take FPC estimates of gas value as our starting point, the three adjustments listed above lead to the gas values shown in Table 11.

In the calculation of construction costs, significant clerical errors were made in the FPC report. One involved the Arctic Gas construction schedule, overstating the present value of cost by $87 million. Another, the failure to increase certain construction items as a consequence of raising the input volume from 2.25 to 2.4 Bcf/d, understated construction costs for Arctic Gas. To correct this error the present value of cost for Arctic Gas must be raised by $15 million. These two corrections require a net reduction in the present value of Arctic Gas construction costs amounting to $72 million.

Table 11

NET NATIONAL ECONOMIC BENEFITS,
ARCTIC GAS, EL PASO, AND ALCAN #2 PROPOSALS
(in millions of 1975 dollars)

	Arctic Gas	El Paso	Alcan #2	Alcan #2 via Dawson
Benefits				
Value of gas	12,929	11,606	12,499	12,462
Value of national security benefit	877	794	854	852
Total benefits	13,806	12,400	13,353	13,314
Costs				
Field gathering and conditioning	1,057	961	1,057	1,057
Field operation and maintenance	41	41	44	44
Transportation facilities	3,790	3,619	4,281	4,640[a]
Working capital	15	34	28	30
System operation and maintenance	304	883	306	319
U.S. other taxes	90	238	198	198
Canadian income taxes	355	—	402	436
Canadian other taxes	32	—	122	132
Total costs	5,684	5,776	6,438	6,856
Net National Economic Benefits	8,122	6,624	6,915	6,458[a]

[a] The cost of transportation facilities assumes that a forty-eight-inch pipeline south of Dawson will be constructed to accommodate 1260 psig pipeline operating pressure. This will be inadequate if a Dawson lateral is constructed and Mackenzie Delta gas is added to 2.4 Bcf/d from Prudhoe Bay. Therefore, the cost of Alcan #2 via Dawson may be understated.

Source: Author's calculations. See text, pp. 75-77.

In addition, because we have delayed the start-up date for the Alcan system, we must stretch out the construction timing for Alcan to correspond more nearly with that of El Paso and Arctic Gas. This adjustment reduces the present value of Alcan's construction costs by $30 million.

The construction cost estimates submitted by Alcan #2 were not as carefully developed as those of the competing lines. The FPC concluded that Alcan's construction costs in Canada were underestimated by at least 10 percent,[17] but the commission did not make the necessary adjustment in the applicant's estimates. The WHTF report noted that consultants to the DOI found that Alcan's costs in Canada were too low by as much as 70 percent.[18] Arctic Gas has pointed out that the Alcan construction cost estimates for Alberta assume that the productivity of labor will be 40 percent higher than Arctic Gas assumed in costing its own system under comparable Canadian conditions.[19] In the face of inadequate support for the Alcan position, the WHTF "decided on a compromise which raises the costs for Alcan within Canada by approximately 30 percent. This amounts to an 11 percent increase in total costs for Alcan. Canadian taxes are also raised by 30 percent to reflect the higher assumed construction costs."[20] Similarly, $491 million of present value has been added to Alcan's estimate of the cost of transportation facilities.[21]

Working capital costs were overestimated for El Paso. The WHTF report noted that late in the preparation of their report it became apparent that the FPC had "included working capital as a cost both in the cost of transportation systems and as a separate cost item in the calculation of net economic benefits for the El Paso system. This overstates the costs of the El Paso system by about $63 million."[22] No correction was made in the WHTF report for this error. We have reduced El Paso construction costs by $34 million of present value to eliminate this double counting of working capital.

Construction costs for Alcan #2 via Dawson require an additional $543.7 million, having a present value of $359 million.

It should be noted that, by Alcan Pipeline Company estimates, a Dempster Highway lateral would increase construction costs 8.5 percent, even though mileage would increase only 4.4 percent. The difference may be accounted for by the greater difficulty of constructing a pipeline via Dawson than near the Alcan highway.

Even this construction cost increase is understated. As indicated earlier, if a Dempster Highway lateral is constructed to ultimately accommodate Mackenzie Delta gas, then the throughput from Dawson

[17] Federal Power Commission, *Recommendation,* p. XIII-19.

[18] White House Task Force, *National Economic Impact,* p. 21.

[19] Arctic Gas, *Submission of the Arctic Gas Project Relative to Selection of an Alaskan Natural Gas Transportation System,* July 1, 1977, p. 55.

[20] White House Task Force, *National Economic Impact,* p. 21.

[21] This estimation is based on Canadian capital costs shown in Alcan Pipeline Company, *Alcan Pipeline Project 48" Alternative Proposal,* March 1977, Section 6, Exhibit 1.

[22] White House Task Force, *National Economic Impact,* p. 21.

south to near the U.S. border would be greater than design capacity. One solution to this foreseeable problem would be to construct the forty-eight-inch pipeline from Dawson to accommodate 1680 psig rather than the planned 1260 psig. This would involve higher construction costs and idle capacity for a decade or more, which would be paid for by gas consumers. These cost increases are not reflected in any of the costs shown in this or any other report.

Again following the WHTF 30 percent compromise on construction costs for the Alcan system, we have raised both Canadian income taxes and "Canadian other taxes" 30 percent.

Both the DOI and the FPC staff reports allowed a credit to Canadian income tax payments by Arctic Gas and Alcan to account for the secondary effects of induced imports from the United States and an income tax generated on profits from the production and sale of these goods and services in the United States. This secondary tax effect was omitted from the FPC recommendations and from the WHTF study. Such secondary effects are generally omitted in cost-benefit analyses. This well-established precedent will be followed here.

Similarly, the DOI analysis included a benefit for energy independence, and this benefit was omitted from the FPC and WHTF reports. The FPC report agreed that the national security benefit was real, but declined to include it on the grounds that the benefit is less tangible than the increased supply of gas and that some of the relationships used to calculate the benefits would be "arbitrary" and "meaningless."[23]

The value of the national security benefit may be calculated as follows: Assuming (1) that delivery of gas from Prudhoe Bay on a regular basis for twenty-five years displaces a need for an equivalent Btu value of crude oil stored under the recently established Strategic Petroleum Reserve System, (2) that the Btu content of one barrel of oil is 5.8 million, (3) that the mid-1975 price of crude oil delivered to the Gulf of Mexico area in the vicinity of salt domes was $12/bbl., (4) that the levelized cost of salt dome storage facilities (capital and operating) is about $0.05/bbl./year, and (5) that the interest cost on capital (at 10 percent) invested in oil is $1.20/bbl. (the total cost is $1.25/bbl./yr. = $0.216/million Btu), then the annual benefit for each of twenty-five years for Prudhoe Bay gas would be as shown in the table at the top of the next page.

It should be pointed out that any energy source capable of displacing oil storage for national security would have the same benefit. In order to credit each project with its potential contribution to this national

---

[23] Federal Power Commission, *Recommendation*, p. I-14.

	Million Btu delivered by each line per year	×	Value of stored oil displaced per million Btu per year	=	Annual value of national security benefit	Mid-1975 value of gas deliveries 1983-2007
Arctic Gas	959,000,000	×	$0.216	=	$207.1 million	$877 million
El Paso	867,800,000*	×	$0.216	=	$187.4 million	$794 million
Alcan #2	933,500,000*	×	$0.216	=	$201.6 million	$854 million
Alcan #2 via Dawson	931,100,000	×	$0.216	=	$201.1 million	$852 million

*Gas delivery data are taken from FPC, *Recommendation to the President,* May 1, 1977, pp. IV-6 and 7. The Arctic Gas deliveries have been adjusted to show a Btu/cu. ft. value of 1168. The Alcan #2 via Dawson route was assigned a shrinkage rate of 6.6 percent. Other data are as shown in the FPC report.

security benefit, the gas delivered by each line has been reduced by its fuel shrinkage. The results are shown in Table 11.

El Paso has stressed the employment impact of its project. While more U.S. employment would be generated by the El Paso project than by either of the others, this is not counted separately as an economic benefit. There is no unemployment problem among the skilled crafts that would be required by El Paso. Unemployment is a persistent problem among the unskilled. The additional demand for skilled labor imposed by El Paso would increase wage inflation in fully employed crafts. Therefore, El Paso's claim of an employment net benefit is rejected in this and all other reports.

Summary of Net National Economic Benefits. A tabulation of NNEB for each pipeline proposal, plus Alcan #2 via Dawson, is provided in Table 11. An Arctic Gas project would clearly earn the greatest economic benefits for the nation, with an NNEB value of $8,122 million. The Alcan #2 project is second best at $6,915 million, and El Paso close behind at $6,624 million. The values for El Paso and Alcan are sufficiently close that, given the margin of error in cost-benefit estimation, the difference is well within the margin of estimating error. The least attractive project from an economic point of view is Alcan #2 via Dawson. The Dawson diversion, suggested by the Canadian National Energy Board as a condition for permit approval, would reduce Alcan's NNEB by $457 million, or 7 percent.

These NNEB values reflect only the social benefits and social costs that have been quantified. Any residual environmental damage not avoided by regulations to be imposed on the construction and operation of a pipeline must be considered separately. The environmental issues have been discussed in Chapter 3. The international political issues discussed in Chapter 4 should also be considered separately.

While the NNEB values shown in Table 11 differ from those developed by the WHTF (which did not evaluate Alcan #2 via Dawson), the rank order of the three proposed pipelines is the same. The FPC report, though it listed Alcan #2 with the highest NNEB, followed closely by Arctic Gas, and El Paso a poor third, commented in the text:

> Alcan's comparative advantage over Arctic occurs for two reasons, both of which are of questionable significance. First, since it appears that Alcan would deliver the gas to market one year earlier, the present value of the total volume of gas delivered will be greater even though the total delivered volumes are almost identical for the two systems. If we were to allow for some increase in the *real* value of the gas over time, a condition that might very well exist, the value of Alcan's earlier delivery would be reduced. Also, as expressed in Chapter VIII, there is good reason to believe Alcan's construction cost estimates may be somewhat low. Thus we cannot find Alcan superior to Arctic on NNEB grounds: Both, however, offer more net benefits than El Paso.[24]

Thus the FPC ratings of the three proposals, like those of the WHTF, are in accord with the ratings shown in this report.

Our NNEB results for the Alcan #2 via Dawson route are of special interest for two reasons. First, this route has not been evaluated in any other analysis. Second, the United States might not have a choice of an Alcan route without the Dawson diversion, in view of the strong position taken by the Canadian National Energy Board. In this event, the net benefits to the United States of an Alcan route are lowered by nearly half a billion dollars and, according to the estimates provided in Table 11, the Alcan route falls to third place, slightly behind El Paso. However, the difference again is well within the range of estimating error. The costs of Alcan #2 via Dawson may be understated depending on how the problem of pipeline capacity in Canada south of Dawson is resolved.

---

[24] Ibid., p. IV-10.

6
THE COST OVERRUN PROBLEM

The Arctic natural gas pipeline project is extremely vulnerable to underestimation of capital costs and subsequent cost overruns. The bidders' game is obvious. Given environmental impacts and national security implications similar to those of its competitors, the proposal with the lowest cost and highest NNEB will probably be the winner. This creates a major incentive to understate true costs. In addition, suppliers of debt and equity capital will insist upon a "cost-of-service" and perhaps also an "all-events" guarantee. The former provision requires that pipeline tariffs be high enough to cover all costs of providing natural gas transportation including debt service and a fair rate of return on equity capital. This is approximately equivalent to a "cost-plus" contract. It reduces the incentive for efficiency since the cost of inefficiency will be fully covered, subject to the usual regulatory commission approval. The all-events provision says that even if the pipeline is started but never completed, equity and debt capital will be returned. Both provisions require that contracts be written to guarantee that rate payers pick up the tab for the full cost of service or for construction costs in the event that a pipeline is started but never completed.

Given the built-in incentive to understate costs, a cost-plus contract leaves rate payers in an extremely vulnerable position. The same situation leaves the nation as a whole vulnerable in the event that cost overruns are so large that net national economic benefits become negative. In this event, the pipeline should never have been built in the first place.

Maximum Acceptable Cost Overrun

The critical question at this point becomes, how large might the

construction-cost overrun become before net national economic benefits decline to zero? In this section a maximum acceptable
construction-cost overrun will be calculated for each project. The relevant data are shown in Table 11 and the calculation is given in Table
12. We find that the Arctic Gas project could sustain the largest cost
overrun. Among the projects that are now politically viable, El Paso
could sustain a 183 percent overrun before its NNEB would disappear.
The net gains to the nation would fall to zero with a 162 percent
overrun in Alcan #2 or with a 139 percent overrun in Alcan #2 via
Dawson.

This analysis is concerned only with a construction-cost overrun
and omits the possibility that operating costs might also be understated
and consequently show cost overruns.

Probable Construction-Cost Overruns

The FPC hearing record is only marginally useful in providing clues to
probable construction-cost overruns. Given the high level of uncertainty surrounding cost estimation, the built-in incentive to underestimate construction costs, and an absence of thorough cost estimates
by objective parties, fears have been expressed that overruns will be
large. One observer of the Alaskan scene wrote, "I would not be
surprised to see the total cost of either project exceed $20 billion."[1]

Presumably each applicant has submitted the lowest capital cost
estimate that will be believable to impartial third parties and will suffer
only minimal damage in cross-examination from competing applicants. The adversary process used in the FPC hearing relies upon the
applicants to challenge the credibility of competing proposals. Accordingly, Arctic Gas has charged that El Paso's capital costs are underestimated and that all of the El Paso facilities combined will incur a 12.6
percent cost overrun. In turn, both El Paso and Arctic Gas charge that
the Alcan proposal is inadequately researched and cannot be constructed under the three-year schedule proposed by Alcan or at the
cost it projects. The FPC staff and administrative law judge appear to
agree with this conclusion. In this section we will attempt to discover
the general range of probable cost overruns for a natural gas pipeline
from the Arctic on the basis of overruns incurred in comparable projects in the past.

Construction-cost overruns in large public and quasi-public construction projects have a long history. Where construction contracts

---

[1] Arlon R. Tussing, letter to the Honorable Chancy Croft, president, Alaska State
Senate, May 26, 1976, p. 3.

Table 12
MAXIMUM CONSTRUCTION-COST OVERRUN
BEFORE NNEB BECOMES ZERO,
ARCTIC GAS, EL PASO, AND ALCAN #2 PROPOSALS

	Arctic Gas	El Paso	Alcan #2	Alcan #2 via Dawson
1. Adjusted NNEB estimates (in millions)	$8,122	$6,624	$6,915	$6,458
2. Present value of construction cost, U.S. share (in millions)	$3,790	$3,619	$4,281	$4,640
3. Sum of lines 1 and 2	$11,912	$10,243	$11,196	$11,098
4. Ratio of line 3 to line 2	3.14	2.83	2.62	2.39
5. Overrun in construction costs which would cause NNEB to become zero	214%	183%	162%	139%
6. Maximum overrun expressed as a compound annual rate, assuming a six-year construction period for all projects[a]	21.0%	18.9%	17.4%	15.6%

[a] The compound annual rate expression is used here only as a convenient method of comparing initial cost estimates with the sum of all actual costs, for several projects independent of construction periods.

Note: This analysis assumes that a construction-cost overrun will have no effect on other costs. Property taxes probably would be higher in proportion to the increase in construction costs. Further, this analysis considers only construction cost overruns and does not include underestimates of operating costs.

Source: Author's calculations based on adjusted NNEB estimates from Table 11.

are based on some form of a cost-plus agreement in which the first step in the process is to secure a permit to begin construction, there is a clear built-in incentive to underestimate capital costs. Bidders on military projects in particular have apparently used modest cost estimates to gain congressional commitments and confronted Congress later with their much higher actual costs. Nonmilitary projects, however, have also registered substantial cost overruns, sometimes for the same reason.

Early research on the cost overrun problem centered on military contracts. These contracts frequently involved a "product" which was not and could not be clearly specified because new research and development were required as part of the contract. Such contracts were commonly of the cost-plus variety. One study based on Department of Defense weapons-development programs in the 1950s concluded as follows: "Development cost predictions made in the face of uncertainty are susceptible to gross errors. For example, the average development cost prediction error in a sample of 12 U.S. weapons programs was 220 percent with a standard deviation of 170 percent."[2] Later research conducted at the Rand Corporation based on weapons-development contracts during the 1960s drew similar conclusions, indicating that Defense Department projects "continue to exhibit an average cost growth of about forty percent (corrected for quality changes and inflation), a schedule slip of about 15 percent, and a final system performance that was likely to deviate 30 to 40 percent from the original specifications."[3] In another Rand Corporation study, Robert Summers attempted to provide a "magic formula" or theory for predicting cost overruns. The results of his econometric analysis indicated that cost overruns were significantly related to the length of time required for the development program and the extent of technological advance involved in the project.[4]

On the assumption that the best estimate of probable construction-cost overruns for an Alaskan gas pipeline should be based on the history of cost overruns for similar projects, the historical record of twelve large construction projects has been examined. They are listed in Table 13. Some of these projects, such as nuclear plants, involved new technologies. Some, such as the Trans-Alaska pipeline (Alyeska), involved old technologies in entirely new environments. None of the twelve is a military project. All have been completed. The construction periods run from three years in the case of Dulles Airport to fourteen years in the case of the Bay Area Rapid Transit System (BART). Initial capital-cost estimates vary from a low of $6.1 million in the case of the Frying Pan, Arkansas, Sugar Loaf Dam Project to $996 million for BART. Overruns vary from 24 percent for the Second Chesapeake Bay Bridge to 755 percent for Alyeska.

In appraising the historical record of construction-cost overruns, adjustments must be made in the initial cost estimate to correct for inflation above the anticipated level included in the initial cost estimate

[2] F. M. Scherer, *The Weapons Acquisition Process: Economic Incentives* (Boston: Harvard University Press, 1961), p. 1.

[3] R. L. Perry, G. K. Smith, A. J. Harman, and S. Henrichsen, *System Acquisition Strategies* (Santa Monica: Rand Corporation, June 1971), R-733-PR/ARPA, p. v.

[4] See R. Summers, *Cost Estimates as Predictors of Actual Weapon Costs: A Study of Major Hardware Articles* (Santa Monica: Rand Corporation, March 1965), RM-3061-PR.

and for changes in the scope of the project. For example, the initial capital-cost estimate for the Rancho Seco Nuclear Power Unit No. 1 constructed by the Sacramento Municipal Utility District was based on a capacity of 800 megawatts. Capacity was increased during construction to 912 megawatts. Similarly, the capacity of Alyeska was increased after the initial estimate of capital cost. The costs of changes in project scope have been estimated and shown in the scope adjustment.

Table 13 shows the unadjusted ratio of final cost to initial cost in column 6. This cost-overrun estimate was then adjusted for each of our two "excused" factors in columns 7 and 8. The adjusted ratio shows that in the case of Alyeska, the final cost was 4.25 times the initial cost estimate. On an adjusted basis, the Alyeska pipeline was 8.6 times the initial estimate. The compound annual rate of cost overrun after adjustments for Alyeska was 22.96 percent.

On an unadjusted basis, the weighted-average completed cost of all twelve projects was 3.93 times the initial estimated capital cost. After adjustment for the two excused factors, the final cost was still 2.21 times the expected cost and the compound annual rate of cost overrun was 10 percent.

The nuclear plants, with their high rates of technological uncertainty, have a relatively high adjusted-cost-overrun rate—8.55 percent compounded annually. The proposed natural gas pipelines are similar to the nuclear plants in that they involve a high level of uncertainty. While gas pipelines have been constructed under permafrost conditions and one oil pipeline has been constructed from Prudhoe Bay to southern Alaska, no forty-eight-inch high pressure pipeline has ever been constructed over great distances within the Arctic Circle (outside of the Soviet Union).

There is another similarity between the nuclear power plants and the proposed gas pipelines. Both involve cost-of-service tariff arrangements and are therefore approximate cost-plus contracts. For this reason, the incentive to minimize actual construction costs is relatively weak.

There is also a point of dissimilarity between the nuclear power plants and the natural gas pipelines. In the case of the power plants, it is not obvious that several applicants competed for the same permit. Instead, the usual procedure is for a single utility or a consortium of utilities to propose a project and seek regulatory approval. In the case of the gas pipeline from the Arctic, three applicants are competing and the rule of the game is clear: the lowest acceptable cost estimate wins a major advantage.

Of the twelve projects considered in Table 13, the Alyeska pipeline is probably most nearly comparable to the natural gas pipeline. The

Table 13

COST OVERRUNS IN MAJOR CONSTRUCTION PROJECTS COMPLETED BETWEEN 1956 AND 1977, ADJUSTED FOR UNANTICIPATED INFLATION AND CHANGES IN PROJECT SCOPE

Project (1)	Initial Estimate — Amount (millions) (2)	Initial Estimate — Date (3)	Actual Result — Amount (millions) (4)
1. Bay Area Rapid Transit Authority	$996.0	1962	$1640.0
2. New Orleans Superdome	46.0	1967	178.0
3. Toledo Edison's Davis-Besse nuclear power plant, Ohio	305.7	1971	466.0
4. Trans-Alaska Oil Pipeline (Alyeska)	900.0[b]	1970	7700.0[c]
5. Cooper Nuclear Station, Nebr. Pub. Power Dist.	184.0	1966	395.3
6. Rancho Seco Nuclear Unit No. 1, Sacramento	142.5	1967	347.0
7. Dulles Airport, Washington, D.C.	66.0[c]	1959	108.3[c]
8. Second Chesapeake Bay Bridge	96.6[c]	1968	120.1[c]
9. Frying Pan Arkansas Project Ruedi Dam	12.8[c]	1962	22.9
10. Frying Pan Arkansas (Sugar Loaf)	6.1[c]	1962[c]	10.2
11. Frying Pan Arkansas (Boustead Tunnel)	9.2[c]	1962	21.2[c]
12. Rayburn Office Building, Washington, D.C.	64.0[c]	1956	98.0[c]

Weighted Average

[a] The compound annual rate expression used in column 9 is used only as a convenient method of comparing initial cost estimates with the sum of all actual costs at the termination of the project. This device permits a comparison of overruns on several projects having different construction periods.

[b] In May 1974, the Alyeska Pipeline Service Co. reestimated capital cost at $4 billion, then in October 1974 costs were again estimated at $6 billion for the completed pipeline. By June 1975, the estimate was raised to $6.375 billion. In 1969, the $900 million cost estimate for

		Ratio after Adjustments		
Date completed (5)	Unadjusted Ratio of Final to Initial Cost (6)	For unanticipated inflation (7)	For change in scope of project (8)	Compound Annual Rate of Cost Overruns, after Adjustments (in percent)[a] (9)
5/76	1.647	1.297	1.037	0.31
7/75	3.870	3.219	3.219	15.73
5/75	1.524	1.401	1.401	11.89
7/77	8.556[c]	6.926	4.250	22.96
74	2.148	1.748	1.748	7.23
74	2.435	2.026	1.239	3.11
62	1.641[c]	1.641[d]	1.486	14.10
6/73	1.243[c]	1.104	1.104	2.00
72	1.789[c]	1.636	1.145	1.36
73	1.672[c]	1.500	1.500	3.75
73	2.304[c]	2.078	1.233	1.92
6/66	1.531[c]	1.531[d]	1.342	2.99
	3.93	3.21	2.21	10.07

Alyeska assumed a capacity of 500 mb/d. The scope was changed to permit a capacity of 1.2 million b/d. The cost of this change in scope was $700 million, raising the initial capital cost estimate to $1.6 billion.

[c] Does not include interest.

[d] Observed inflation was less than anticipated.

Source: Details of this cost overrun analysis are available from the author, on request.

environmental similarity is obvious. Two points of dissimilarity, however, are also present and important. First, cost underestimation in the Alyeska case was not a result of competition among multiple applicants each attempting to show least cost. Second, an economic incentive existed in the Alyeska case to keep costs under control since the pipeline equity risk was taken by the owners of the oil resources. The "net back" of revenues to the oil companies will be reduced by any and all cost overruns. In spite of this important economic incentive, the Alyeska overrun rate was nearly 23 percent compounded annually after adjustments for inflation and an expansion in scope.

If tariffs covering the transportation of gas through an Alaskan pipeline are governed by a cost-of-service provision, then this cost-plus feature might be expected to generate an overrun as great as that of Alyeska. But a 23 percent overrun rate would generate a total cost 3.46 times the estimated construction cost. Line 4 of Table 12 shows that an overrun of this magnitude would eliminate all of the net national economic benefits from any proposed gas pipeline. A 10 percent cost overrun corresponding to the average for all twelve projects studied would cause the final construction cost to exceed the initial estimate by a factor of 1.77, well within the maximum acceptable overrun as computed in Table 12.

This analysis of the cost overrun record of other large construction projects cannot determine the probable overrun on a specific Alaskan natural gas project. At best, the analysis merely suggests the extent of the problem to be anticipated. It serves as a warning for gas consumers and for regulatory authorities. The extent of the cost overrun will probably be determined by a combination of the unique risks of each project and the incentive system which guides management in the construction phase. If an all-events, cost-of-service tariff is granted, then the project will approximate a cost-plus contract and the incentive to economize on construction costs will simply not be present. This leaves natural gas consumers extremely vulnerable.

Project-Specific Estimates of Cost Overruns

As a supplement to this general analysis, project-specific risk and overrun analysis can shed additional light on the magnitude of probable cost overruns. Risk analysis conducted by the DOI was made prior to the submission of the Alcan #2 proposal and was therefore concerned primarily with the Arctic Gas and El Paso schemes. This risk analysis examined the possibility of schedule slippage and consequent cost overruns. Consideration was given to construction risks, engineering problems, labor disputes, supply delays, manageability problems, and weather delays. The DOI analysis concludes that

> It does not seem unreasonable that the El Paso system can slip six to eighteen months, probably due to difficulties in bringing the liquefication process on line, and that the Arctic Gas system can slip twelve to thirty-six months, most probably because of compounding difficulties with Arctic winter construction, logistical support, and labor delays on the Canadian section.[5]

The DOI staff roughly assessed the magnitude of probable schedule slippage at "one year for the El Paso system and two years for the Arctic Gas system," which would add approximately $1 billion in cost overrun to the El Paso proposal and $2 billion to the Arctic Gas system.[6]

The risk analysis completed by the FPC staff reached conclusions contrary to those of the DOI report. The FPC staff held that

> because of a heavier reliance on new techniques and on an unconventional type of transportation mode, it appears that the El Paso project would be much more susceptible to inordinate cost overruns than the other two proposals. Unlike the other proposals, El Paso's more southerly transportation mode is the much less traditional cryogenic tanker fleet. Furthermore, the design of both the LNG plant and tankers involves a technological advance, though probably sound, that may subject the project to a greater degree of possible overruns.[7]

The FPC staff further concluded that "the degree of certainty that can be attached to the cost projections associated with the Alcan project is significantly less than for the other two projects. . . ."[8]

The White House task force concerned with construction delays and cost overruns reviewed the DOI and FPC risk analyses and performed its own analysis including a brief study of the Alcan #2 project. The task force concluded that delays greater than those foreseen by the other two agencies should be expected. Further, it rated the Arctic Gas project as the highest risk and El Paso the lowest, as indicated in Table 14.

There are major discrepancies between the task force overrun

---

[5] United States Department of the Interior, *Alaskan Natural Gas Transportation Systems,* December 1975, p. 143.

[6] Ibid.

[7] Federal Power Commission, "Position Brief of the Commission Staff," December 7, 1976, p. 19.

[8] Ibid., p. 7.

Table 14

WHITE HOUSE TASK FORCE ESTIMATES OF
CONSTRUCTION DELAYS AND COST OVERRUNS,
ARCTIC GAS, EL PASO, AND ALCAN #2 PROPOSALS

Applicant	Expected Schedule Delay (in months)[a]	Expected Construction-Cost Overrun (in percent)[b]
Arctic Gas	20	37
El Paso	15	31
Alcan #2	17	32

[a] The expected value is the mean or "average" value of the estimated probability distribution.

[b] The construction-cost overrun estimates do not include interest cost during construction.

Source: White House Task Force, *Construction Delay and Cost Overruns*, July 1, 1977, p. 2.

estimates and the actual overruns incurred in the twelve cases reported in Table 13. Because the task force estimates are independent of inflation and changes in scope, the comparable figures for actual overrun are those from which the effects of inflation and scope changes have been removed—shown in column 8 of Table 13. We find that the average historical overrun was 121 percent. In contrast, the task force forecasts overruns in the range of 32-37 percent. Of the twelve projects studied, only two had overruns below the range expected by the task force, while ten had overruns above the top of the range. This comparison suggests that the task force seriously underestimated the probable construction-cost overruns.

Nevertheless, applying the task force cost-overrun estimates to the construction schedule for each project, we have computed the present value of such cost overruns and deducted those present value losses from each project's NNEB. The calculations are presented in Table 15. The results show the following: (1) There is no change in the rating of the three projects—Arctic Gas is still the economically preferred route, and El Paso is last. (2) El Paso and Alcan #2 are even closer together than by other measures and have essentially identical NNEB values. (3) All three routes have large NNEB values.

Nowhere in the risk analysis conducted by DOI, its contractor (the Aerospace Corporation), the FPC, or the White House task force was any consideration given to the risk of interruption in service due to a

Table 15

EFFECT OF COST OVERRUNS ESTIMATED BY
WHITE HOUSE TASK FORCE ON NNEB

(in millions of 1975 dollars)

Applicant	NNEB before Cost Overrun	Present Value of Cost Overrun	NNEB after Cost Overrun
Arctic Gas	8,122	1,270	6,852
El Paso	6,624	977	5,647
Alcan #2	6,915	1,190	5,725

Source: Author's calculations.

labor dispute once the project is in service. Yet there is a substantial difference between the all-pipeline routes and the El Paso proposal in this respect. A conventional buried pipeline is capital intensive. In the event of a strike, the essential services of a pipeline can be maintained by supervisory and management personnel. Therefore, a strike would not disrupt normal gas flows to users.

The operation of the El Paso system, involving LNG conversion and transportation, is more labor intensive. It is very unlikely that any of the fleet of LNG carriers could be operated by supervisory and management personnel. Therefore, a strike by a maritime union would shut down gas supplies from the North Slope and reduce total gas supplies to the lower forty-eight states by as much as 5 percent. This could create a significant economic disruption corresponding closely in percentage supply reduction with the oil supply curtailment during the 1973-1974 oil embargo. A no-strike pledge is under discussion for inclusion in any maritime contract. However, such provisions are hard to enforce. Massive "sickness" on the part of organized workers has the same effect as a strike.

Summary of Cost-Overrun Analysis

In the case of Arctic Gas, construction costs might rise as much as 3.14 times the initial estimate before NNEB for this project would decline to zero. For El Paso, construction costs are low relative to operating costs. Even though El Paso does not have the second highest NNEB, it could sustain the second highest cost overrun without exhausting its net national benefits. Its limit would be reached when construction costs rose to 2.83 times the initial estimate. Alcan #2 and Alcan #2 via Dawson could sustain construction costs as high as 2.62 and 2.39 times the initial estimates before their respective NNEBs would become zero.

An examination of the record of twelve large construction projects that have been completed within the last few years shows that cost overruns of this magnitude are not uncommon. The average cost overrun showed final costs 2.21 times initial estimates, after adjustments for inflation and scope changes. The final cost of Alyeska was 4.25 times the initial estimate. Thus, the record shows that the limits given by zero NNEB can easily be exceeded. Given the built-in incentive for applicants to underestimate costs and their requests for an all events, cost-of-service tariff, gas consumers and the public are vulnerable to losses in real economic values.

Project-specific overrun forecasts prepared by DOI, FPC, and the WHTF differ by project but do not change the NNEB ranking of the projects and do indicate that, even with the overruns expected by the WHTF study, any of the three projects would produce large scale net economic benefits for the nation. However, the agencies' forecasts of probable overruns are well below the cost overruns actually incurred by similar projects and appear to underestimate probable overruns.

7
RATE REGULATION
AND FINANCING

The Unique Problems of Rate Regulation and Financing

The construction of a natural gas transportation system from the Arctic poses two unique problems for rate regulation and financing. First, any of the proposed pipelines would be larger than any gas pipeline ever constructed. While the Alyeska oil pipeline will end up costing about the same as the combined U.S. and Canadian shares of the Arctic Gas proposal, the *projected* cost of Alyeska was only $0.9 billion. The *projected* costs for the gas pipelines range from $5 to $7 billion.

Second, relative to other public-utility capital investments, a gas pipeline from the Arctic involves inordinately high risks. For the Alcan proposal, 42 percent of the pipeline within Alaska and Canada is located north of the Arctic Circle. This is extremely high-risk construction. In the case of El Paso, 37 percent of the proposed pipeline is above the Arctic Circle. The El Paso proposal also involves high-risk LNG plants and LNG ship construction. High risk translates into the possibility of large construction-cost overruns.

These two unique problems have, in turn, spawned three subproblems. First, in the event of great unforeseen construction difficulties producing large cost overruns, the system might never be completed. While this possibility appears to be remote, its probability is greater than zero. Should this occur, regulation must specify how the debt and equity capital losses are to be divided.

Second, in the highly probable event that the pipeline is completed and placed into service but suffers one or more prolonged periods of interruption, how is the cost of interruption to be divided between consumers and financiers, and what are the rights of debt holders for their debt service relative to those of equity holders for their rates of return on equity? Prolonged interruption appears to be a greater problem for the El Paso route where the probability of a major accident in the LNG facility, in one or more tankers, and in the regasification facility is high relative to the probability of accident for a conventional buried pipeline. Further, the repair time from a serious accident in these El Paso facilities is likely to be long compared with that in a conventional pipeline.

Third, how are pipeline tariffs to be determined such that debt repayment and service are clearly provided and the rate of return on equity capital conforms with legal requirements? Given the enormous debt and equity funding required for any of these projects, debt capital is not likely to flow unless legally binding and workable provision is made for debt repayment and service in all events. Similarly, in a regulatory environment where rates of return on equity cannot legally exceed a "fair rate of return," equity capital will not flow if there is a possibility that any adversity will result in a loss of equity capital or failure to earn the expected rate of return.

The nature of the financial risk has been identified succinctly in an El Paso brief as follows:

> The prospect that the project's revenues may be insufficient to permit it to make the necessary payments arises from three risks: (1) failure of governmental regulation to permit the project to generate sufficient revenues; (2) prolonged interruption of the project's operation without receipt of revenues; (3) noncompletion of the project for economic, physical or other reasons.[1]

Positions of the Parties

The Applicants. While the applicants differ on their recommendations regarding rate regulation and financing, the critical points in their positions are virtually identical. All applicants require an all-events, cost-of-service tariff.

The first element in this regulatory system would provide that the regulatory agencies be bound to order tariffs which would protect both debt and equity-capital suppliers in all events. These events include (1) noncompletion of the pipeline for whatever reason, and (2) completion of the pipeline, but prolonged interruption of service.

The cost-of-service element of this proposed regulatory system would provide that all necessary and normal costs of pipeline construction, operation, and financing be fully covered in pipeline tariffs. This provision is intended to include construction cost overruns, without limit. The El Paso-Alaska Company has presented testimony by L. E.

---

[1] El Paso-Alaska Company, "Brief of the El Paso-Alaska Company Concerning Project Financing and Regulatory Action," June 4, 1976, p. 2.

Katzenbach of the investment firm, White, Weld and Company, pointing out that both investors and lenders will require assurances from the FPC that

> all operating and financial costs incurred by the owner-operator including provisions for payment of these costs without regard to completion of the project, interruption of service, or *force majeure*. The equity investors will insist upon a tariff which will provide an equitable return on their investment, due consideration being given to the venture risk taken.[2]

In addition to the all-events, cost-of-service tariff, Alcan believes that a U.S. Treasury "backstop" is needed to provide additional financing in the event of a major cost overrun.[3] El Paso denies the need for Treasury backstop financing for its project. To the extent that such backstop financing is needed, it is likely to involve a "lender of last resort" role in which the government would provide the completion assurance needed by lenders. This assurance could take the form of either loans or guarantees of loans to complete the project once started. In addition, backstopping might provide that the lender of last resort would pay off all indebtedness incurred to the point of project abortion. In the event of prolonged interruption, an insurer of last resort might be required to pay tariff charges of the project companies, after exhaustion of any private business interruption insurance, or to pay off all indebtedness then outstanding and discontinue the project.

The FPC. As a matter of principle, the Federal Power Commission staff is in agreement with the applicants on the cost-of-service rate-making system. However, the FPC staff specifically rejects applicant requests for " 'carte blanche' authority to pass through *any* costs to the shippers without any prior determination of justness and reasonableness of those costs."[4] Instead, the staff position is the customary one which

---

[2] El Paso-Alaska Company, "Reply Brief of El Paso-Alaska Company on Tariff Matters," December 15, 1976, p. 1.

[3] The Alcan #1 proposal commingled Canadian and American gas, making private financing unworkable, thus requiring both Canadian and U.S. government backstopping. The Alcan #2 proposal involved transporting only U.S. gas, and Alcan held that, given certain specific assumptions, U.S. government guarantees would not be needed. Then with the NEB action setting up contingent plans for tapping Mackenzie Delta gas via a Dempster lateral, again commingling Canadian and U.S. gas, plus the NEB decision that no Canadian taxpayer support would be available, the Alcan project again became nonfinanceable in the absence of U.S. government backstopping. See White, Weld and Company, "Impact of the NEB Decision on Alcan's Ability to Finance," July 25, 1977.

[4] Federal Power Commission, "Initial Tariff Brief of the Commission Staff," May 28, 1976, p. 3.

"allows" operating costs which pass FPC scrutiny. The difference between these two positions may in fact be trivial. Historically, regulatory commissions rarely disallow major capital or operating expense outlays.

Similarly, the FPC staff rejects a guarantee that in the event of noncompletion, costs in all events would be passed on to consumers. However, the staff wrote that "the Commission cannot preclude flow-through of any cost element that meets the [Natural Gas Act] standards, and in that way the lenders are assured of an appropriate judicial determination of the flow-through issue. They can require no more than that."[5]

The Treasury Department. The U.S. Treasury Department has also offered testimony regarding the regulation and finance problem. This testimony strongly endorses private financing of the project and strongly opposes any residual burden being placed on taxpayers through any backstopping role for the government. This testimony points out that the principal beneficiaries of gas production from the Arctic are gas consumers. In addition, owners of gas resources, including both the companies involved and the state of Alaska, have a substantial beneficiary interest. The state of Alaska is likely to receive a flow of revenue amounting to about $650 million per year when Prudhoe Bay gas begins to flow at its full level. The private companies will benefit by receiving a wellhead price of gas in excess of their full cost of production. In the view of the Treasury Department, the financial base for this project should be broadened beyond the sponsoring companies. Equity investment should be solicited from both the producer companies and the state of Alaska. Residual risks should be borne by these groups and by consumer beneficiaries, not by taxpayers in general, according to the Treasury.

The Dilemma of Regulation and Finance

Running throughout the formal hearings concerned with alternative pipeline projects one finds a basic dilemma arising out of the enormous capital requirements of the project and the inordinately high risk. Suppliers of both debt and equity capital insist upon an all-events, cost-of-service tariff. But this is a cost-plus pricing system and as such leaves an inadequate incentive for efficient construction of the pipeline. The greater the costs of construction, the higher the rate base

---

[5] Ibid., p. 17.

on which a return to the pipeline firm is calculated. The higher the rate base, the greater the absolute return to equity holders. Thus, an all-events, cost-of-service tariff almost guarantees a large construction-cost overrun. This is one horn of the dilemma.

The other horn of the dilemma is that, if an all-events, cost-of-service tariff is denied, neither debt nor equity funding will flow to the project.

The FPC solution to this dilemma is to grant a cost-of-service tariff but reject the all-events element of the applicant proposal, then permit companies to petition for compensation in the event of major loss. The FPC would attempt to eliminate inefficiency arising out of the cost-of-service element of the tariff by attaching a condition to the applicant's certificates "giving the FPC Staff the right to audit the books of the Applicants during the construction period."[6] Testimony by the regulatory staff of the FPC indicated that a considerable auditing effort would be necessary to insure that costs were properly recorded in accordance with the commission's Uniform System of Accounts.[7]

There are several reasons why this FPC audit approach is an inadequate means of protecting the public against cost overruns and gross inefficiency. First, the FPC staff is not in a position to identify inefficiency. One must be very close to the management of a project and be well trained and experienced in business decision-making in order to pass judgment on whether too many employees are assigned to a given task, to evaluate the kind of machinery purchased and used, to separate normal from abnormal accidents and damage, to evaluate expensive precautions and the like. Second, any judgment rendered by the commission staff which would disallow some expenses as unjust or unreasonable after the fact would do nothing to avoid the social costs already incurred. All that the FPC could do after the fact would be to redistribute the private costs between the rate base to be passed on to consumers and a charge against equity to be absorbed by the suppliers of equity capital. Third, the historical record of regulatory commission disallowance of costs indicates that, with insignificant exceptions, costs incurred are found to be just and reasonable. For these reasons, efficiency is not likely to be achieved through FPC auditing.

The problems of cost-of-service pricing under conditions in which large overruns are possible are compounded by the practice of "rolled-in" pricing. This procedure averages high-cost new supplies with low-cost old production under price controls, usually resulting in an

---

[6] Ibid., p. 14.

[7] Ibid., p. 15.

average price far below the cost of new resources. This leads to re-source misallocation. Users continue to consume a scarce resource like natural gas as if it were a cheap and renewable product, when in fact new supplies bear a social cost well above the average price.

The problem of rolled-in pricing is easily illustrated by the situation at hand. Under the present FPC price control system, the average price of all gas purchased by major interstate pipelines in the United States as of October 1976 was $0.69/Mcf.[8] If the average cost of new gas supplies from the lowest cost source, which might be Prudhoe Bay gas, is $1.74/Mcf (see Table 10 showing the minimum possible cost of gas delivered by Alcan via Dawson), and this source supplies an additional 5 percent of the delivered gas, then the rolled-in price would be only $0.74/Mcf. This average price would encourage consumers to use a scarce nonrenewable resource which has a real incremental cost of $1.74/Mcf. Efficient resource allocation would require that market prices reflect the real cost of new resources. In the example just given, if prices corresponded with the cost of new supplies, consumers would be led to economize on this scarce nonrenewable resource. Rolled-in pricing permits the pricing authority to hide high costs. This practice deceives consumers and consequently misallocates scarce resources.

The Missing Link

The missing link in the regulation and financing deliberations has been a lack of incentive in two areas. First, there is no incentive to achieve construction efficiency. If a pipeline is constructed below estimated capital costs, the tariff system established by the FPC will grant only a fair rate of return on the reduced capital investment. On the other hand, with an all-events, cost-of-service tariff, great construction inefficiency will result in large overruns. Historically, these costs have gone into the rate base and a fair rate of return has been granted on this larger rate base.

Second, throughout the FPC hearing, concern was expressed about whether capital would flow into an Arctic pipeline given the enormity of the capital requirement and risk. As an economic problem, the answer to this often posed question is relatively simple. Capital will always flow into investments that are deemed to be attractive and will avoid those that are not. As El Paso witness L. E. Katzenbach testified, "The United States long-term capital market is the broadest, most sophisticated and, generally, lowest-cost market in the world."[9] The

---

[8] Federal Energy Administration, *Monthly Energy Review,* April 1977, p. 76.

[9] "Comparison of the Proposed Financing Plans under the Alaskan Gas Transportation Projects," June 4, 1976, p. 7.

capital market is no different from any other competitive market in which resources flow if the price is right.

The pipeline transportation sector has historically been a regulated sector, and no one who testified in the hearing proposed that it be removed from the category of "natural monopolies" subject to regulation. The terms of regulation are one of the critical elements in the flow of capital. The other critical element is the basic profitability of the project. This latter issue has been discussed above.

Proposal #1. Given the basic dilemma—that in the absence of an all-events, cost-of-service tariff capital will not flow, and that with such a tariff and its cost-plus characteristic, the resulting inefficiency and rate burden to be placed on consumers may be enormous—are there more attractive alternatives?

Two proposals will be offered here. They are designed to avoid the dilemma identified above. The first involves an efficiency incentive system for equity capital suppliers and a solution to the unwillingness of creditors to provide debt money in the absence of the offensive tariff guarantee.

Debt capital will flow to a gas pipeline project at low interest rates if, and only if, the perceived risk is minimal. It is proposed that the U.S. Treasury Department guarantee, for a fee, the entire debt capital required for the U.S. share of a northern Alaskan gas transportation system. This fee should be set on a commercial basis as far as possible. The fee would compensate the government for the risk it assumes. With this guarantee, capital would flow in the desired amount at the lowest possible rate of interest. The necessary cost would be passed on to the ultimate consumers and not to the taxpayers, in accordance with the Treasury's views. The fee structure might have two parts: (1) a relatively low fee would be charged on the debt capital needs as projected in the hearings, and (2) a higher fee would be assessed for any overrun debt capital.

Equity capital will flow into a gas transportation system if the rate of return is deemed, by the market, to be attractive. Treasury witness J. M. Niehuss testified that a relatively high rate of return was justified by the unique nature of this project and the extent of the commitment that the equity investment alone represented to most participating shippers.[10] The applicants have requested that a rate of return in the range of 15 to 17 percent after taxes be allowed on this project.

It is proposed that a rate of return at the top of this range (17

---

[10] Department of the Treasury, "Staff Brief of the Department of the Treasury," December 1976, pp. 6-7.

Figure 6

PROPOSAL #1: RATE-OF-RETURN MODEL

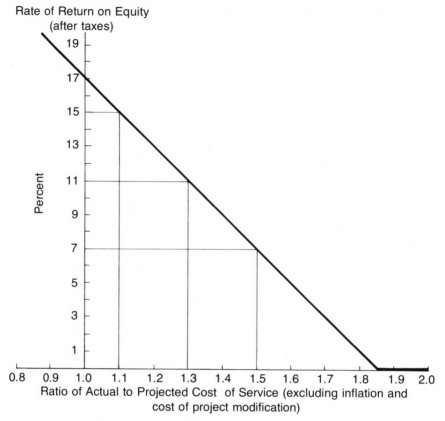

Rate of Return on Equity (after taxes)

Percent

Ratio of Actual to Projected Cost of Service (excluding inflation and cost of project modification)

Note: For an outline of Proposal #1, see text, pp. 101-105.
Source: Author.

percent after taxes) on equity be allowed if, and only if, the applicant winning the permit does in fact supply natural gas at the cost of service which that application has proposed. As a powerful incentive to stimulate construction efficiency, a further premium rate of return would be permitted if the system were constructed at a capital cost below the projected cost with the result that the cost of service was below the applicant's own projection. Conversely, substantial cost overruns would result in the cost of service being greater than projected. In this event, lower rates of return would be allowed.

This proposal is shown graphically in Figure 6. This model is only illustrative since rates of return must reflect current money-market and risk conditions. It suggests that if the ratio of the actual to the projected cost of service is 1.0, then a 17 percent after-tax rate of return would be granted.

A construction-cost overrun that resulted in a ratio of actual to projected cost of service amounting to 1.3 would provide a rate of return after taxes of only 11 percent. In the event that cost overruns were of such magnitude that the actual cost of service exceeded the projected cost by 85 percent or more, then the rate of return to the equity holders would be zero.[11] On the other end of the spectrum, the model suggests that if construction efficiency permitted a cost-of-service charge of 10 percent below that projected, a premium of 19 percent return would be awarded.

A regulatory system based on a rate of return that was a function of the ratio of actual to projected cost of service would serve the public interest in that it would provide a strong incentive for efficient construction and operation. In the model, a 10 percent saving in cost of service due to efficient management would cost only 2 percent on the equity base.

The penalties for the equity owners proposed here are possible because there are compensating potential rewards at the other end of the efficiency scale. Under the present regulatory system there is no practical way to penalize *inefficiency*. This follows from the fact that the reward for *efficiency* is limited to the fair rate of return. If the FPC threatens, through its proposed audit system, to disallow capital outlays that are deemed to be inefficient (unreasonable costs) the commission may not be taken seriously; if it is, equity capital will simply not flow into the project.

This proposal would also result in greater efficiency in regulation. Surveillance of construction by the FPC would become unnecessary. The rate-of-return incentive would be far more powerful than any audit and, in terms of past history, much more effective.

It should be made clear that both actual and projected costs must be calculated in terms of the same dollars. This means that the effect of inflation must be factored out of the calculated actual cost of service. The degree of inflation in the economy is a function of monetary policy and is not subject to any degree of control by the builders of a gas transportation system. Similarly, any change in the scope of the project must be factored out. Thus, if a government decision is made to transport either more or less gas, the projected and actual costs must be

---

[11] A negative rate of return is not included in the model. Negative rates would probably not be permitted by the courts on the grounds that they would constitute a confiscation of capital.

adjusted accordingly. All three applicants have provided a variety of cost estimates corresponding to different volumes of gas to be transported.

This proposal would obviate the need for both the all-events and the cost-of-service tariff requests. The debt holder's position would be fully guaranteed upon payment of the guarantee fee to the Treasury. Hence, debt holders would not need the requested protection. Equity capital suppliers would be given the opportunity to earn a high rate of return as a reward for efficiency. This reward would be the compensation for accepting the risks of equity capital.

One might object that the proposal would be difficult to administer. This objection would be well taken. However, the present system too is difficult to administer. Under the present system the regulatory agency must determine what costs are to be allowed into the rate base. It must determine regularly what operating costs are acceptable as just and reasonable and therefore deductible from revenues. The agency must determine what rates charged to shippers will generate the desired rate of return to equity holders. Finally, the agency must determine what constitutes a fair rate of return. These are difficult problems. The only advantage of the existing system is that it has been adjudicated over many years and the rules of the game are reasonably well understood by both regulators and regulated. The fact remains, however, that the system is basically a cost-plus system. The primary goal of regulation historically has been to ensure that the regulated utility does not earn more than a fair rate of return. The system has never successfully implemented an efficiency goal, only a rate-of-return goal.

The proposed system would require new legislation. Under current statute, a fair rate of return is mandated. This does not permit a premium above the fair rate of return as a reward for efficiency, nor does it permit a lower rate as a penalty for inefficiency. Further, legislation would be required to authorize a Treasury Department debt guarantee contract in exchange for a guarantee fee.

This proposal might also be rejected by the applicant companies. It is unlikely, however, that they would reject it outright. What is more likely is that negotiations might take place concerning the rate-of-return schedule. This negotiation itself would be of great value. It would smoke out any underestimation of capital and operating costs in the applicants' proposals. Under the present system, these underestimates will remain hidden until the nation's resources are committed to the pipeline project. At that point, the project company will apply to the regulatory agency for approval of a tariff based on its actual capital cost in accordance with the established cost-of-service principle.

Arctic Gas wrote that the "sine qua non of any financing plan for any transportation system to transport gas from Arctic areas to markets is a total security arrangement package, satisfactory to lenders, that shippers will accept, in their capacity as both shippers and as equity investors." [12] The essential ingredients of the financial plan listed by Arctic Gas or the other applicants do not include a reference to the interest of consumers in the efficient transportation of gas. Similarly, the FPC staff, in arguing for its favored approach, wrote that "the cost-of-service rate, when properly defined, meets all of the requisites of the Natural Gas Act and will permit the successful Applicants to maintain a healthy economic position for rendering its jurisdictional service. Further, such a rate will *assure* the consumers of receiving this valuable service at the lowest reasonable cost." [13] There is no evidence in the history of regulatory practice to support the contention that rate regulation will "assure" consumers of low and reasonable costs. Efficient construction and operation are not the issue in regulation. The issue is a fair rate of return. Low and reasonable cost of service requires efficient construction and operation, as well as a fair rate of return to the suppliers of equity capital.

Proposal #2. There is a second attractive solution to the regulatory dilemma. This solution addresses the efficiency-incentive problem, but not the issue of financial backstopping. However, it would reduce the latter problem in scope.

Two economic groups—the Prudhoe Bay operating companies and the state of Alaska—those that are now producing oil from the North Slope and would produce gas—have a strong economic interest in the efficient construction and operation of a gas pipeline. Both are residual claimants. Their income from gas ownership and production is a function of (1) the delivered price of gas and (2) the cost of production and pipeline transportation. Any and all construction-cost and operating overruns will reduce their net income. These strong economic incentives for efficiency can be captured by letting those two groups construct a gas pipeline using any route approved by the governments of Canada and the United States.

To implement this incentive system, the FPC should deny all present applications and await a proposal (or proposals) by the operating companies and the state of Alaska. [14] If the expected private benefits exceed the costs, resulting in attractive expected profits, a proposal

---

[12] Arctic Gas, "Brief of the Arctic Gas Project Relative to Financing Brief," December 13, 1976, p. 31.
[13] Federal Power Commission, "Initial Tariff Brief," p. 3.

will be forthcoming. Cost-of-service and minimum-delivered-price estimates shown in Chapter 6 indicate that the construction of a pipeline would be profitable in the absence of government control over Prudhoe Bay wellhead prices. Because of the apparent private profitability under these specific conditions, no government subsidy would be necessary to "internalize" the national security benefit. Given the presence of a foreign tax, this private (but not social) cost could be corrected by the power of the FPC and the U.S. government to approve or reject proposals.

A further advantage of this proposal, in addition to the fact that it would create incentives for efficiency, is its simplicity. There would be no need for a regulatory agency to audit decisions about construction. The incentive to build and operate the pipeline efficiently would lead the ownership interests to institute and administer cost controls far more effectively than could be expected of a regulatory agency.

An objection may be raised that this system did not appear to work well in the Alyeska case where a large overrun was recorded. The only answer possible is that without the residual claimant's interest, and with public regulation as the primary custodian and enforcer of efficient construction, costs would have been even higher. The long delay in government approval of Alyeska and the extremely expensive and inadequately foreseen environmental protection measures imposed on the oil pipeline were unique factors causing cost overruns in Alyeska. These factors need not recur in a gas pipeline construction project.

One possible disadvantage of this second proposal is that it raises the issue of joint venture ownership of a pipeline by the same companies that own the resource. However, this was not deemed to be a significant issue in the parallel situation of Alyeska and it poses no new problem for a gas pipeline.

The financing problem would be moderated if this proposal were adopted. Relative to the gas pipeline sponsors, the Prudhoe Bay oil companies are credit worthy and could raise the necessary debt funding, as they did in financing Alyeska. Further, this credit worthiness will be improved for both the oil companies and the state of Alaska as large cash flows occur from Prudhoe Bay oil production.

---

[14] Following the same line of reasoning, the U.S. Treasury Department offered a similar proposal calling for equity financing by these two groups, in addition to gas transmission and distribution companies. See "Supplemental Testimony of John M. Niehuss," November 8, 1976, pp. 2-5. I am indebted to Professor P. E. Sorensen of Florida State University for assistance in refining this proposal.

8
CONCLUSIONS AND RECOMMENDATIONS

(1) On the basis of quantified net national economic benefits and cost-of-service data, we find that any of the proposed projects would yield large gains to the nation. The relevant data are as follows:

	NNEB (in millions of 1975 dollars)	Minimum Cost of Gas Delivered to Retail Distributors in the Lower 48 States (in 1975 dollars/Mcf)
Arctic Gas	8,122	1.49
El Paso	6,624	1.77
Alcan #2	6,915	1.70
Alcan #2 via Dawson	6,458	1.74

The Arctic Gas project would yield the greatest NNEB and would deliver gas to retail distributors in the lower forty-eight states at the least cost. But the recent Canadian Berger Commission and National Energy Board recommendations and decisions, followed by the decision of eight U. S. sponsors to shift their support to Alcan #2, appear to have removed Arctic Gas as a viable alternative for the United States.

Among the remaining options, Alcan #2 would yield slightly higher NNEBs and would deliver gas to lower forty-eight state markets at lower cost than would the El Paso project. However, the Canadian National Energy Board may require that the Alcan route be diverted to run through Dawson in the Yukon Territory. In this event, the slight advantage formerly held by Alcan #2 over El Paso would be lost. For all practical purposes, NNEB values and minimum delivered costs for the two routes would become identical. The observed differences are easily explainable in terms of estimating errors.

(2) Since NNEB and delivered cost appear to be roughly equal for the El Paso and Alcan proposals, the choice of the optimum project should be based on the nonquantified variables. From an environmental point of view, the analysis provided in Chapter 3 indicates a clear preference for the Alcan route, with or without the Dawson diversion. From an international political point of view, the analysis in Chapter 4 indicates that a trans-Canada route is acceptable, if the cost-benefit analysis indicates a preference for this route.

The Canadian National Energy Board recommended that the Alcan route be assessed up to $200 million to pay for social-economic costs in the territorial areas north of the 60th parallel. This recommendation was reaffirmed by the Lysyk report. Any such charges would be fully borne by gas consumers in the United States. Before any U.S. approval of a trans-Canada route, firm commitments should be obtained concerning such charges. An additional $200 million payable in 1982 would have a 1975 present value of $100 million. The Dawson diversion that might be required as a condition for a Canadian government permit would add another $450 million of costs (present value). These two items would cause the NNEB for El Paso to exceed that of Alcan by nearly $400 million of present value. While we have no quantified estimate of the net environmental benefits of Alcan, it appears unlikely that the net present value of those environmental benefits would exceed $400 million. This means that the United States has no strong reason to prefer either of the two routes.

One other nonquantified factor is the possibility of El Paso service interruptions due to a strike by the maritime union involved in LNG transportation. The bargaining power of a maritime union able to halt deliveries for an extended period of time and the relatively labor intensive nature of the El Paso project become significant penalties for El Paso and lead to a preference for Alcan.

(3) The possibility of major cost overruns was extensively analyzed. Given the implicit decision-making structure where each applicant has an incentive to understate his expected capital and operating costs, significant cost overruns must be anticipated. The record of twelve similar projects shows an average cost overrun (after adjustments for inflation and changes in project scope) of 121 percent. Both El Paso and Alcan could sustain an overrun of this magnitude and still yield significant net economic benefits to the nation. However, if such overruns occurred, consumers would pay higher prices for gas, and/or owners and producers of Prudhoe Bay gas would receive less

for their gas. Neither project would show a positive NNEB value if it incurred an overrun comparable to Alyeska's 325 percent.

The project-specific overrun estimates by government agencies appear to have underestimated the probable cost overruns. Nevertheless, these estimates are helpful in differentiating the overrun risks of the alternative projects. The White House task force indicated that Alcan would have both a greater construction schedule delay and a larger construction-cost overrun than El Paso, but the differences are not large.

The decision-making structure virtually guarantees cost overruns for two reasons: First, applicants are competing for government permits to build a gas transportation system and the lowest cost proposal has an advantage. Second, the regulatory system is essentially a cost-plus system. Transportation rates will be set so as to yield a fair rate of return on the investment, as determined by a regulatory commission. All applicants demand an all-events, cost-of-service tariff in which cost overruns lead to virtually guaranteed higher tariff rates for gas transportation. In essence, this is equivalent to a cost-plus contract. The incentive to construct and operate a transportation system efficiently is relatively weak.

(4) Given weak incentives to efficient construction and operation, large overruns must be expected. Two alternative proposals have been offered which are designed to change the regulatory system and build in efficiency incentives.

- One proposal would make the allowed rate of return on invested capital vary inversely with the extent of the construction-cost overrun, as reflected in the required cost of service. A rate of return on equity at the top of the range requested by the applicants (17 percent after taxes) would be allowed if, and only if, the applicant winning the permit supplied natural gas at the cost of service which that applicant had proposed. Further, an additional premium return would be allowed, according to an ascending schedule, if the cost of service were below that projected by the applicant. Conversely, in the event of cost overruns and cost of service higher than claimed, lower rates of return would be set, on a descending scale.

This system would provide a powerful incentive for the efficient construction and operation of a gas transportation system. Further, it would reduce the cost of regulation because surveillance of costs in the construction process would be unnecessary. Adjustments in the initial-capital-cost estimate would be allowed for two factors only: price inflation and changes in project scope.

Debt capital would be guaranteed by the U.S. Treasury, for a guarantee fee, set on the basis of commercial principles. Acceptance of this proposal would obviate the need for the objectionable all-events, cost-of-service tariff.

This system would smoke out any known underestimates in the construction plan. It would provide a powerful incentive to avoid cost overruns that would be passed on to consumers. It would contribute to a more nearly optimal allocation of scarce resources.

• The second proposal has the virtue of greater simplicity. There are two economic groups that have a strong interest in the construction and operation of a highly efficient transportation system, namely the Prudhoe Bay operating companies and the state of Alaska. The returns to both would automatically decline as transportation costs increased. This proposal would require denial of all present applications. A proposal would then be invited with these two groups as joint sponsors and equity-capital suppliers. The governments of the United States and Canada would still select the route. Data on system costs and revenues clearly indicate that, in the absence of cost overruns and government control over Prudhoe Bay wellhead gas prices, gas production and transportation would be deemed profitable.

These two groups have a stronger incentive to construct and operate a transportation system with greater efficiency than could be enforced by any conceivable regulatory system. Further, the cost of regulation would be reduced sharply because efficiency incentives would be built in.

(5) Finally, before a decision is made by either government, extensive intergovernmental negotiations should be held. From a narrow U.S. point of view, while the Alcan #2 route is preferred, its margin of superiority is small. Both countries would be better off with Alcan than with the El Paso route. This conclusion follows from the fact that the present value of income tax payments to the Canadian government, at the expense of U.S. gas consumers, would amount to $402 million. While this must be viewed as a social cost to the United States, it is a transfer payment from the combined U.S.-Canadian viewpoint. Canada has an interest in an Alcan route that might ultimately reduce the cost of delivering Mackenzie Delta gas to Canadian markets. If it is true that both Canada and the United States would gain by an Alcan route, then negotiation should be capable of reaching a "division of the pie" advantageous to both. The same argument applies with added force to the Arctic Gas proposal, apparently rejected by Canada.

Negotiation may lead to extended delay, but the benefits of delay are likely to exceed the costs. Delay would permit (1) more careful delineation of the Mackenzie Delta reserves, (2) further exploration of the Beaufort Sea reserves, (3) seismic (and possibly other) exploration of the large gas resources that the U.S. Geological Survey forecasts for the Arctic National Wildlife Range, (4) more careful cost and environmental research connected with the inadequately researched Alcan #2 route, and (5) resolution by Canada of the native land claims issue. Further, if the value of natural gas increases faster than the discount rate used to evaluate alternatives, then extended delay in construction would be socially desirable in promoting conservation of resources.

Cover and book design: Pat Taylor